GALILEO'S
PENDULUM

GALILEO'S PENDULUM

From the Rhythm of Time
to the Making of Matter

ROGER G. NEWTON

HARVARD UNIVERSITY PRESS

Cambridge, Massachusetts
London, England
2004

Library of Congress Cataloging-in-Publication Data
Newton, Roger G.
Galileo's pendulum : from the rhythm of time to the making of matter /
Roger G. Newton.
p. cm.
Includes bibliographical references and index.
ISBN 0-674-01331-X (alk. paper)
1. Time measurements.   2. Pendulum.   I. Title.
QB209.N48 2004
529'.7—dc22
2003056972

To Benjamin, whose time has just begun

Contents

Preface

This book is about both ancient history and modern science. Though its subject is not Greek mythology, its aim may be freely interpreted as describing the pervasive influence of Terpsichore, the Muse of dance and rhythm, over our lives, and how science, through Galileo's pendulum, managed to tame that which Euripides called "measureless and wild"—time. The periodic motion of that swinging bob exerted a remarkable influence on the development of modern science and mathematics. This influence eventually extended over not only our understanding of all the natural phenomena that vary with time but even the way science views the very existence of the material world—the particles that make up the objects we touch as well as the light and sound with which we sense our surroundings. The Greek myth may thus be seen as completed when Terpsichore, with the aid of science, finally deconstructed her father, the powerful Zeus, master of the universe.

I am indebted to two biologists, David Kehoe and Arthur Koch, for discussions and advice on matters in their field of which, as a physicist, I was ignorant. If I still did not get some things quite right, it is entirely my fault and not theirs. My debt also extends to my wife, Ruth, for invaluable editorial assistance.

GALILEO'S
PENDULUM

Introduction

He was seventeen and bored listening to the Mass being celebrated in the cathedral of Pisa. Looking for some object to arrest his attention, the young medical student began to focus on a chandelier high above his head, hanging from a long, thin chain, swinging gently to and fro in the spring breeze. How long does it take for the oscillations to repeat themselves, he wondered, timing them with his pulse. To his astonishment, he found that the lamp took as many pulse beats to complete a swing when hardly moving at all as when the wind made it sway more widely. The name of the perceptive young man, destined to make other momentous scientific discoveries, was Galileo Galilei.

The legend of how Galileo discovered the *isochronism* of the simple pendulum, as told by his biographer Vincenzio Viviani, is apocryphal, but neither the fact that he found it nor the profound effects it had on our civilization in later centuries can be denied. This book is about the rhythm of time, how that rhythm was finally regulated by Galileo's pendulum, the impact the oscillations of the pendulum had on our perception of that rhythm, and how these oscillations were later found to manifest themselves in many other natural phenomena.

The first three chapters set the stage, describing the rhythms of time as registered before the stabilizing advent of the pendulum swing: the imprinting of the succession of night and day on living organisms, the history of the calendar, characterized by the struggle of civilizations to reconcile the lunar and solar cycles, and the reckoning of shorter time intervals until the Middle Ages. Neither the biological mechanisms that nature employed to implant rhythms in living creatures nor the methods human cultures developed to keep track of the periods imposed on us by the heavens were stable or accurate. But the former conferred a sufficient adaptive advantage to make them a ubiquitous feature of life, and for eons the latter fulfilled people's needs adequately.

After the Renaissance, however, commercial and scientific advances exerted an urgent pressure to measure time more accurately, and the further progress of Western civilization would have been severely hampered without the invention of a stable and steady clock. Large-scale maritime navigation as well as modern science depended on it. The pendulum and, later, other physically equivalent mechanisms served this purpose admirably.

Surprisingly, the physics of the *harmonic oscillator*—that is, Galileo's pendulum—which made it possible to regulate the flow of time, leads far beyond a mere device for making accurate clocks. These oscillators have been found to be the basis not only of what we hear as the sound of music and see as the colors of light but, via the quantum theory, of what we understand as the fabric of the universe. Without oscillators, there would be no particles: no air to breathe, no fluids to sustain life, and no solid matter to form the earth. Here is the story of the simplest, yet most fundamental physical system in nature and how it ties the rhythm of time together with our very material existence.

1

Biological Timekeeping:
The Body's Rhythms

Imagine living on a large chunk of rock hurtling endlessly through space. It revolves neither about a central sun nor on its own axis, and it has no satellite circling it. (Never mind that such a world would lack light and heat to sustain life.) There are no mornings, no evenings, no summers, no winters, only a monotonous, temporally undifferentiated world.

Would intelligent creatures, had they evolved there, have either internal clocks or a concept of the flow of time? The answer may well be no. Our entire notion of the passage of time is based on the perception of periodic change, on the repeated transition from day to night—the rising and setting of the sun marking the daytime, of the moon visible at night—and on the recurrence of the seasons. Time was born of rhythm, and periodicity is part of who we are.

Flying his *Winnie Mae* westward around the world in 1931, the American aviator Wiley Post is reputed to have been the first to explore and record that bothersome phenomenon we are now all familiar with, jet lag. Observing the effects of time shifts from one longitude to another on the performance of pilots, he is said to have found that the human body cannot adjust itself to changing time zones faster than about 2 hours per day. How much of his experience was simply the effect of fatigue, however, remains unclear.

Although biological mechanisms do not work with the accuracy or stability of modern clocks, a sense of time and its rhythms is built into the functioning of the human body. Our heart, with its beating pulse, is the clock-like internal rhythm we are most aware of. In his discovery of the law of the pendulum, which turned out to have the most profound effect on all later time-measuring devices, Galileo used—if legend can be believed—his own pulse beat as the test. There are, however, other biological timekeepers with longer periods that also play important roles in our lives. While their coordinated rates of progress are autonomous overall, we know that after a few days our inner clocks can be reset and will fall in step with a shifted rhythm; our lack of synchronization with the local time, even after a long flight across the Atlantic or the Pacific, slowly disappears.

The technical term, introduced in 1959, for the internal timer that keeps track of this 24-hour periodicity and retains it even in the absence of external cues is the *circadian* system (from the Latin *circa* for "about" or "approximately" and *dies* for "day" because its period is approximately one day). Though more or less known to biologists for over two hundred years, this and other biological clocks have been the subject of intensive research during the last half century.

The first human physiological variables that scientists observed to be governed by a circadian rhythm were pulse rate and body temperature. Even while resting in bed and fasting, the deep body temperature of a person varies by almost one degree centigrade between its low in the early morning hours and a high late in the afternoon. As discovered more than 150 years ago, the excretions of our kidneys, which regulate the concentrations of vital substances such as salt in our blood, rise and fall in the course of the day. More than 100 additional physiological and psychological variables are also subject to diurnal periodicities. For example, the speed with which children can do computations varies by about 10 percent between its slowest value in the early morning to a high before noon, dropping to a nadir in the early afternoon, rising again to a peak about 6 o'clock in the afternoon, and then falling off in the evening. This was first measured in 1907 and again a half century later, and both studies came up with similar results.

The extremely controversial question that arose immediately was to what extent this human circadian rhythm was an autonomous mechanism rather than a simple response to external signals such as changes in the level of light, the times of meals, or social interactions with our surroundings. It has not been easy to find the answer. In some cases the suspected extraneous cues, such as possible intruding electric fields that varied in the course of a day, were very subtle, but careful experiments on both human and nonhuman subjects at many scientific laboratories have led to the definite conclusion that our body contains an autonomous timekeeper. Individuals who volunteered to be kept in total artificial isolation for extended periods, sometimes shut up in caves with no time cues of any kind, helped find the answers. In 1938 two researchers from the University of Chicago lived in a chamber in Mammoth Cave, Kentucky, for 32 days; some thirty years later, a French speleologist spent two months in a cold cave, 375 feet under ground in the Alps. The Frenchman called his above-ground supporters by telephone whenever he ate, went to sleep, and woke, and he recorded in detail his thoughts and impressions of the passage of time. All such explorers found themselves subject to definite internal time signals. It turned out, however, that the measured period of their bodily variables (all of which were consistent with one another), as well as their subjective impression of the time of day, their periods of sleep and waking, were slightly longer than 25 hours; by the time they emerged from their prolonged isolation, their internal timer was many hours out of phase with the external 24-hour clock.

Two issues here have to be clearly distinguished: the *rate* at which the biological clock advances (which determines the *period* of its rhythm), and the *phase* of its rhythm at a given time. Even an accurate watch worn on a flight from New York to Los Angeles will be three hours *out of phase* upon arrival and will have to be reset. A watch that runs slow—that is, whose rate of advance lags—will become more and more out of phase with accurate timekeepers as time goes by; because its period is longer than that of an accurate reference clock, only repeated resetting will make it show the correct time.

In some instances the unrecognized existence of these circadian rhythms has interfered with experiments set up to measure psychological stimulus-response reactions that would not be expected to vary with time. What is more, because the endogenous rhythm differs somewhat from 24 hours, the interference was not always easy to detect: even repeating an experiment at the same time every day did not necessarily eliminate a systematic error introduced by variations in a subject's reaction depending on the time of day when the test was performed. When more than one rhythm is at work—the external diurnal and the slightly different circadian—the result can be quite confusing.

The autonomy of biological clocks is now a well-established fact. In a person isolated from environmental cues, the human circadian system proceeds at a fairly uniform speed that is longer than the 24-hour day. In a person who is not isolated, it is kept synchronous with the cycle of the sun through constant entraining in response to environmental variations—mostly variations in light intensity. In other words, though running at a steady rate, our internal clock is slow by about an hour per day, but since it is continually automatically reset by cycles of light and dark, under normal circumstances the loss of time is not cumulative; our internal clock is thus entrained with the rhythm of the sun.

Humans, of course, are only one species, and the circadian is only one of the biological rhythms. In addition to the heart beat, some internal timekeepers have shorter periods—they are called *ultradian*—such as the 0.1 second period of the electric activity of the brain measured in an electroencephalogram (EEG). Still others are longer, such as the 28-day menstrual cycle of women and the *circannual* timekeepers regulating hibernation in bears; possibly some internal clocks have much longer periods. Rhythmic behavior appears to be a property of most biological systems.

The synchronized flashing of fireflies displays a short period (about 1 second but varying from species to species). Male fireflies in Malaysia and New Guinea congregate by the thousands in trees, flashing rhythmically and in unison, producing a spectacular display. Their ultradian periods of bright light emission are found to be self-sustaining and endogenous, but when they congregate in large groups they mutually entrain their flashing to be synchronous. (Entrainment can also be initiated by a blinking flashlight.)[1]

In some cases, a behavior that appears to depend on a long-term timing device may actually require only a circadian system. The Canadian snowshoe rabbit, for example, whose pelt is brown in the summer and white in the winter, starts changing the color of its coat well before the first snowfall, to take advantage of camouflage. It recognizes when August comes around not by a circannual clock but by relying on its circadian timer to keep track of the number of hours of daylight as compared with darkness. If blindfolded for part of the day in July, it will alter its appearance earlier. In other mammals, circannual as well as circadian clocks regulate metabolism and reproductive cycles so that enough fat is deposited for the winter and young are born at a season of abundant food supply and favorable weather. In these instances, too, the length of the day helps to entrain the rhythm, even though that rhythm has a much longer period. Some crabs living in coastal waters have *circa-tidal* clocks that keep track of the environmental rhythm most vital for them, the tides. There are other maritime animals whose spawning and fertility are governed by internal timers geared directly to the cycle of the moon.

Serious research on circadian timing systems began after

a serendipitous observation by the Swiss physician August Forel in 1910. Eating his breakfast on the terrace of his house in the Alps one morning shortly after his arrival for the summer, he noticed that the open jam jar on the table attracted a few bees from a nearby hive. On subsequent mornings, however, he was intrigued to see them come just before breakfast was ready, as though they anticipated the arrival of the jam. When he then decided to avoid the pesky critters by eating inside, they still came at the appointed time to the table outside, looking for food. Since the bees did not visit his house at any other time of day, he concluded they must have a "memory of time."

Some twenty years later, the ethologists Ingeborg Beling and Karl von Frisch conducted systematic observations of individually marked bees that had been offered sugar water for several days at a fixed place and time. On the day of observation, when no food was presented, the researchers recorded the arrival of each bee and found a sharp peak during the "training" time (with a slight preference for early arrival). However, this time-training was successful only when the interval between feedings was 24 hours, or close to it; it failed for intervals of 19 or 48 hours between feedings. To make sure that the bees were not sensing some external cues that the experimenters had missed, Beling conducted some of his experiments in a salt mine, deep under ground. Max Renner trained bees in his laboratory in France to expect food at 8:15 P.M. but tested them the next day after a transatlantic flight to New York City. There, they looked for their snacks as usual, but at about 8:15 P.M. French time.

Hamsters and other rodents turned out to be particularly useful for measuring circadian rhythms in mammals. Kept under steady illumination, without any indications of the time of day, in a cage with an exercise wheel in which they could run at will, they would exercise regularly every day at about the same time for very similar periods; their feeding and drinking schedules were found to be equally regular. Such studies have shown that, without any external cues, many circadian "habits" persist for very long periods, even for the life of the animal; others begin to fade earlier.

Until the middle of the twentieth century, the remarkable ability of migratory birds to navigate over long distances was regarded as a great mystery. Now we know that it is based at least in part on their circadian system. The Pacific golden plover manages to find its way over 2,000 miles of ocean from the Aleutians to Hawaii in the fall and back in the spring. An internal clock allows these birds to be guided by the sun, even though its position changes during the course of a day: in the morning, flying south means keeping the sun on their left; in the afternoon, on the right. The zoologist Gustav Kramer was the first to demonstrate that birds are able to keep track of where the sun should be, depending on the time of day. He observed that during the migrating season most starlings in his circular aviary, which was lit by the sun through slits in the walls, would perch facing the direction in which they would have migrated if free. When, by means of mirrors, he suddenly changed the direction of illumination, the birds would change their direction of perch accordingly, thereby showing that they use the sun as their guide. But when the direction of the sun's illumination changed very gradually in the course of the day, it had no effect on them; their circadian clock allowed them to adjust for the sun's movement.

Other migrating birds were found to be able to navigate

not only by the sun but even by the stars at night, again by using their built-in timer. That's how whitethroated warblers navigate their flight annually in the fall from northern Europe across the Adriatic Sea and the Mediterranean to Egypt and back in the spring; during cloudy nights they become confused. Certain species of ants, too, have been discovered to find their way every day to the same foraging spot by using the direction of the sun in conjunction with a circadian clock. If enclosed in a dark box, they lose their orientation, but upon release into the sunny open, they quickly find their previous direction and continue on their way, even though the sun has meanwhile moved.

Internal timing systems exist in plants as well. That many plants react to the time of day by opening or closing their leaves or blossoms has been known for a long time. Androsthenes, one of Alexander the Great's generals on his march to India, recorded that the leaves of the tamarind trees he saw there opened in the morning and folded up at night. Most gardeners have noticed that some flowers close up their petals in the evening and reopen them at dawn or later, while the daylily opens a new bloom every morning. The great eighteenth-century botanist Carolus Linnaeus designed a timekeeping flowerbed, with specially arranged species of flowers whose blossoms would open at specific times of the day (see Figure 1). The circadian timer in bees presumably coevolved to enable them, without the need for external cues from a visible sun, to visit their favorite flowers for nectar at times when their blossoms were open, so as not to waste energy on unnecessary scouting.

Is such behavior in plants a simple response to the waxing and waning daylight or is it caused by an internal mech-

Eine Blumen-Uhr

1. A picture of the flower clock designed by Linnaeus in 1751, with the characteristic times of petal opening and closing of various species.

anism built into the plant's structure? To find the answer, the French astronomer Jean-Jacques De Mairan performed experiments on mimosa trees, which, like the Indian tamarind tree, fold up their leaves in the evening and reopen them in the morning. In 1729 he reported his findings to the French Royal Academy: the mimosa did not require the assistance of the daily cycle of changing light; even in total darkness (he kept them shut up in a dark room for lengthy periods of time) it would open its leaves and fold them more or less at the same time as the onset of day and night, respectively. While its rhythm was not entirely independent of the external diurnal cycle, the plant did keep, at least for a while, an imprint of the daily change of light with which it matured. During the nineteenth century, the study of the daily leaf movements of plants was taken up by the botanists Francis Darwin (third son of Charles), William Pfeffer, and Julius von Sachs (the latter two in Germany), as well as their students and collaborators.

For many years the circadian internal clock of plants, like that of animals, was considered quite mysterious, but after much research over the last half century, pioneered by the German biologist Erwin Bünning, its existence is now well established and many of its components have been isolated. Like animals, plants have endogenous cycles of various durations. For example, as though in preparation for the coming winter, many plants seem to have an internal circannual timer that enables them to anticipate the onset of the cold season and harden their cells so as to survive a sudden frost. As in the case of the snowshoe rabbit, often this apparently circannual timing is, in reality, a response to the length of day, measured by means of a circadian clock. These internal rhythms are either inherited or learned from external stimuli, but they persist for various lengths of time after the latter are removed, and their periods do not vary with ambient temperature. In order to remain synchronous with the outside world, however, the internal clock has to be entrained and regularly reset by light cues.

Though rhythmic temperature changes also work in some instances, the cues that serve to entrain most biological clocks—these are called zeitgebers (timegivers) by physiologists—seem to be varying levels of light intensity. This is hardly surprising, as the origin of the diurnal rhythm of our life is the experience of the succession of day by night; that environmental fact must have been of key importance in the evolution of the internal timing mechanism. The process of entrainment is not necessarily a sudden, discontinuous resetting of the clock, however. In many instances it is, after an instantaneous initiation, a gradual change from an old phase to a new one, like the slow turning of the hands of a clock by an external agent, without altering its intrinsic period, that is, without causing it to go faster or slower from that point on.[2]

Remarkably enough, circadian rhythms exist even in the simplest biological organisms, such as the single-celled blue-green algae (cyanobacteria), in which an internal clock paces the 24-hour cycles of nitrogen fixation and photosynthesis. The cycles serve to separate these two processes in time, which is crucial for their proper functioning because the presence of oxygen—a product of photosynthesis—interferes with nitrogen fixation. The marine dinoflagellate *Gonyaulax polyedra* is a single-celled bioluminescent alga that exists in very large quantities in coastal waters off the West Coast of North America. Bioluminescence is a chemical process in which an enzyme, luciferase, oxidizes a luciferin molecule, producing a new molecule at a higher energy state, which subsequently descends to its ground state, emitting a light quantum. *Gonyaulax* exhibits two kinds of luminescence: a low-intensity steady glow that varies rhythmically, peaking near the approach of dawn, and a much brighter rhythmic flash, at its maximum near midnight, that requires an external disturbance to set it off. In addition, *Gonyaulax* also has other circadian rhythms, including one which causes a sharp spike in cell division at dawn, and another spike in its photosynthetic capacity. If cultivated in extremely dim continuous light, photosynthesis will proceed at a steady, nonrhythmic rate; but under normal conditions it has a distinct diurnal periodicity, maintaining all of its rhythms for weeks when under steady, dim illumination. In total darkness, with its photosynthesis at a standstill, it quickly loses all its rhythms but regains them at their previous pace when photosynthesis is restored by a short, intense flash of light.

Surprising as it may seem, this simple organism, consisting of a single cell, can thus sustain many independent circadian and other rhythms. Indeed, it has been demonstrated that the timers contained in single cells govern even the biological periodicities of large, complicated organs such as the human heart, whose ongoing rhythm keeps us alive. Some forty years ago, Isaac Harary at the University of California in Los Angeles invented a process to separate the cells of a rat's heart and keep them alive in a nutrient medium. Examining them under a microscope, he found that some of these single cells were still beating at the old rhythm, all on their own. What is perhaps most surprising is that even some single-celled bacteria that divide and multiply once every 16 hours manage to pass along identical copies of their 24-hour clocks, running and in phase, to their off-spring! Almost every gene in these organisms is controlled by their circadian system.

In many birds, circadian rhythms appear to be controlled by the pineal gland, which Descartes thought was the seat of the soul in humans. This photosensitive organ rhythmically produces the hormone melatonin, which signals the gland's activity to other parts of the body, even in vitro. When cultured in an environment with 12 hours of light and 12 hours of darkness and then transferred to one with constant illumination, pineal tissue will continue emitting cyclically varying amounts of melatonin for several days. What is more, when pineal tissue was transplanted from one bird to another bird whose pineal gland had been removed, the new host not only regained the circadian rhythms it had previously lost, but the phase of its newly acquired rhythms was that of the donor: the host acted as if it lived in the donor's time zone. Even pieces of the gland in vitro will for some time continue to produce melatonin at a rhythmic rate.

On the other hand, in mammals, including humans, the primary central pacemakers of most, though not all, of these rhythms are the two clusters of nerve cell bodies called the *suprachiasmatic nuclei* (SCN), located in the hypothalamus just above the optic chiasm behind the retina. The SCN are able to generate rhythms on their own, without any kind of light cues. Newborn rats, for example, kept in total darkness, exhibit a distinct circadian rhythm even on their first day of life. Under normal operations, however, the SCN are synchronized with the day-night cycle of their surroundings, the entrainment paths being visual projections directly from the retina. In fact, it has very recently been discovered that the eye contains special photo sensors, not used for vision, connected directly to the SCN, enabling even some people who are blind to entrain their circadian rhythm.[3] In addition, the hypothalamus also contains another, distinct ultradian pacemaker that regulates the release of a follicle-stimulating hormone every hour—a *circhoral* clock. Many separate circadian oscillators have now been found scattered throughout the body, to regulate local rhythms in behavior and physiological functioning. These timers can maintain their rhythms by themselves for only a few days and are constantly entrained by signals from the master clock, the SCN.[4]

The question of whether the rhythms of the many kinds of biological clocks found in organisms are imprinted on them by external signals early in life or, instead, are genetically determined has been answered by detailed experiments, particularly those performed on the ubiquitous fruit fly. *Drosophila melanogaster* has two relatively stable circadian systems, one regulating its rest-activity cycle and another governing the times of day at which their young emerge from their pupae. In 1979 R. J. Konopka produced mutants that had no periodicity at all, others that had short periods of about 19 hours, and still others that had periods as long as 28 hours. Each mutant differed by a single gene (thereafter named *per*) from the standard laboratory fruit fly (the "wild type"). Geneticists also studied the fungus *Neurospora,* whose production of spores is governed by a circadian system. Mutants exhibited a range of different periods; in a mixture of cells with various mutants as well as wild types, the length of the circadian period turned out to be proportional to the percentage of nuclei with mutant genes. In this case, there was no unique genetic locus that determined the length of the circadian period in a population.[5]

What is the mechanism that makes these clocks tick? That chemistry alone is capable of producing rhythmic behavior has been known since 1950, when the Russian chemist Boris Belousov discovered what has come to be called the Belousov-Zhabotinsky reaction, which produces rhythmically changing waves that separate reaction products of different colors (see Figure 2).[6] In plants, the clock mechanism has to be of an analogous purely biochemical nature. Indeed, experiments performed on the energy metabolism of yeast cells fermenting in suspension have shown that when the metabolic steady state is perturbed by the addition of a substrate, their fermenting energy production—called glycolysis—begins to oscillate with a period of several minutes, and these variations in glycolysis can then be studied in vitro in a cell-free preparation, with nothing at work to engender them but chemistry.

The observed oscillatory behavior is the typical consequence of what engineers call a feedback mechanism: the enzyme doing the glycolysis, converting fructose-6-phosphate (F-6-P) to fructose-diphosphate (FDP), is itself activated by FDP. Thus, when the glycolysis begins to produce FDP, it is stimulated and accelerates, diminishing the concentration of the original F-6-P, thereby inhibiting further FDP production and slowing the fermentation until, after a considerable delay, the F-6-P has been replenished by newly arriving stock. At the right concentration and metabolic flow, the levels of both F-6-P and FDP will, as a result, change rhythmically, each with the same period but out of

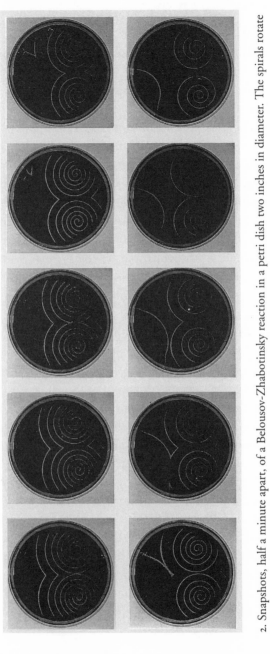

2. Snapshots, half a minute apart, of a Belousov-Zhabotinsky reaction in a petri dish two inches in diameter. The spirals rotate once per minute.

phase, with F-6-P low when FDP is high, and vice versa. (The delay caused by the slowness of the biochemical processes involved is crucial in producing the rhythmic changes in concentration rather than a quick relaxation into equilibrium, as it would without the time lag.) Choosing appropriate enzymes, concentrations, and flow rates in a system without any cells, biologists have been able to achieve metabolic oscillations with 24-hour periods—circadian rhythms by purely chemical means. A similar feedback mechanism has been found to produce the circadian rhythm in *Drosophila,* only in this case the places of F-6-P and FDP are taken by the PER protein produced by the *per* gene and the messenger RNA it synthesizes, as well as several other proteins. (Though the principles of the mechanism are the same, the proteins involved in producing the biological clock in the cyanobacteria are quite different from those in *Drosophila,* from which we may conclude that circadian systems are unlikely to have evolved from a single common ancestor.)

The outcome of such feedback is often described by a *limit cycle,* in which the plot of the amount of one of the reagents against that of the other approaches the same closed curve more and more closely, no matter where they started out (see Figure 3). The concentrations of the two substances then perform a dance, each varying from high to low with a fixed rhythm.

The mechanism responsible for generating the cyclic behavior of a much more structured and complicated organ such as the SCN is not yet well understood and remains an important subject for continuing research. Is the rhythmicity the product of a network of synapses or nerve cells act-

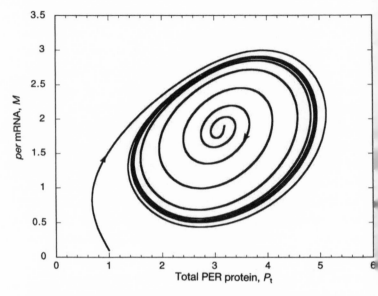

3. Limit cycle model of sustained oscillations in PER protein and *per* mRNA in *Drosophila*. The plot of one against the other gradually approaches a closed curve (shown darker), either from the inside or from the outside.

ing collectively, or is it the result of the action of pacemakers in individual cells, as in plants? There are indications that the latter is the case. In vitro, even individual neurons of the SCN retain their circadian rhythms for many days running, but collaborative effects may be involved as well. As mentioned earlier, there are also partially autonomous circadian pacemakers in the body which are constantly entrained by the SCN, but the signaling mechanisms are still not fully known. One thing, however, can be stated with certainty: even though the literature frequently refers to the endogenous rhythmic behavior of many biological systems as "oscillatory," no physical oscillators are present in any of them, which is the fundamental reason why these living clocks are able to tell time only *circa* and never with complete accuracy. There would, however, have been no adaptive advantage for organisms to produce a clock of greater stability, even if that had been possible. Mechanisms requiring daily entrainment have served quite satisfactorily for eons.

In the next chapter, we will turn outward, away from biological systems to the various elaborate designs that human cultures have invented to record the passage of time, from the calendar to the watch with a second hand. Long after unreliable timing devices had been in use for millennia, the pendulum was discovered to be an unfailing instrument when the need for real stability and accuracy in the measuring of time finally arose.

2

The Calendar: Different Drummers

Whether early humans were consciously aware of the continuous flow of time is of course impossible to tell, but it appears that this notion was firmly in place by the time civilization arose. Early civilizations had three good reasons to make use of the rhythmic nature of their experience of time: to measure the duration of a given process, such as the length of travel time from one place to another, to calculate how long ago certain memorable events took place, and to specify the *now*, as well as a future *now*, for ordering their lives and their relations to others.

For the first purpose, simple comparisons were initially sufficient: your voyage will take twice as long as it takes to travel from Shiraz to Isphahan. But as civilization advanced, it became necessary to divide a time interval into several equal parts. When sailors on board ship or soldiers at an army camp had to take turns at night manning the watch, such duties were surely expected to be of more or less equal duration. For the second purpose, early civilizations employed the simple counting of universally recognized recurring events: the disaster happened 10 floodings of the Nile ago; we have had three new moons since your father left.

The third requirement answered such questions as: Is now the right time to sow corn or to plant beans? When is the appropriate time to plow the fields?

The need for telling the time of year must have arisen first among hunter-gatherers. When should we store provisions for the coming winter? What is the best season to hunt bison or deer? With the advent of agriculture, however, plowing, planting, and harvesting had to occur at the proper times, and it was no accident that the calendar was invented in those civilizations where farming first evolved. As the need for such arcane knowledge of the proper seasons increased, so did the power and mystery of those who were able to provide it. Since the sun, moon, and stars had already been imbued with religious and superstitious significance for millennia, early astronomy and the construction of a calendar based on it were intimately associated with the practice of religion, and the keepers and purveyors of the mysteriously acquired information were the priests, who jealously guarded their sources. The construction of Stonehenge some 3,500 years ago thus served both a religious and a practical purpose.[1]

Even as the agricultural need for a calendar advanced the power of the priests, so the religious requirements of placating the gods and showing them proper respect by observing a regular schedule of feast days further strengthened the desire to have a rigidly ordered calendar. After all, the movements of the heavenly bodies appeared to be the only regular and predictable activities of the otherwise capricious and willful gods. Thus, astrology developed into astronomy, which became the incubator of mathematics. (Under Roman law, astrologers were called "the mathematicians.")

The one periodic phenomenon in nature that required no experts to observe and measure was the succession of night and day. Initially regarded as separate units, the light and dark periods were eventually combined into a single entity, the full day. It began at sunrise in some cultures, at dusk in others. A longer natural unit in the rhythm of human life was the cycle of the moon: the time from one full moon to the next measured a period of approximately 30 days. Finally, the year, with its repetition of the seasons, became the third periodic unit noticeable to everyone, and the impetus for the almanac.

The rhythms of the light-dark cycle and the seasons are, of course, governed by the sun. The former originates from the rotation of the earth about its axis (facing, and then facing away from, the sun), while the latter arises from the earth's orbit about the sun. Because the earth's axis is tilted relative to the plane of its orbit, the northern hemisphere inclines away from the sun in one section of the orbit, while the southern hemisphere inclines toward the sun, yielding winter in the north and summer in the south; in the opposite section, it is summer in the north and winter in the south. The period of the earth's revolution around the sun is not a whole-number multiple of its rotation time: it takes the earth 365.25 days (rotations; equal to roughly 12 lunar months) to complete one orbit. All the complications of establishing a well-functioning yearly calendar arose from the need to reconcile these three periods: the day, the month, and the year.

Since 12 lunar months fall short of a solar year, it became necessary to insert some extra time into the lunar calendar to keep it in sync with the length of a solar year—a process called *intercalation*. Various cultures and civilizations han-

dled this problem in different ways. Many primitive socie-
ties considered the months of deepest winter useless and
ignored them. The Central Eskimos, for example, had a
"sunless month" in the winter, which was of indeterminate
length and was dropped during the years when the new
moon coincided with the winter solstice.

Marco Polo reported in the thirteenth century CE that for
all their important occasions the Mongols consulted an almanac supposedly adopted from a Chinese system going
back as far as the third millennium BCE. The Chinese practice in very ancient times, however, was to leave the construction of calendars up to local communities. This lack of
uniformity led to the introduction of an official calendar,
under the Three Dynasties in the second millennium BCE.
In later times, every emperor was anxious to improve on the
efforts of the last, and during the two millennia after ca. 370
BCE they constructed about 100 new, increasingly accurate
almanacs, showing not only the months and days but also
the motions of the moon, the sun, and the planets. These
calendars were based on a lunar cycle with 12 "Earthly
Branches" and a number of supplementary days called
"Heavenly Stems" required to reconcile the periods of the
sun and the moon. So important did the emperors regard
their almanacs that by the fourteenth century CE the government printed 3 million copies of them—they included
information about lucky and unlucky days, best days for
travel, weddings, and business transactions, and so on—as
official documents, which no one else was permitted to
copy. The Chinese did not number their years but called
them (along with the months, the days, and even the hours)
by names such as Rat, Ox, Tiger, Hare, Dragon, Serpent,
Horse, Sheep, Ape, Cock, Dog, and Swine.

Few details seem to be known about the Sumerian calendar except that the days were grouped into weeks of seven days, based on their astrology. Sumerian practices were lost for a long time, but the seven-day week became more or less universal—the Jews adopted it from the Assyrians, and Christians from the Jews—with some notable recent exceptions. Driven by a desire to weaken the influence of the Catholic Church after the French Revolution, a committee in France consisting of the mathematicians Pierre-Simon de Laplace, Joseph-Louis Lagrange, Gaspard Monge, and the poet Fabre d'Églantine introduced a decimal calendar with a 10-day décade and a month consisting of 3 décades. This lasted only 13 years. For similar anti-Church reasons, the USSR in 1929 introduced a month consisting of 6 weeks, each of 5 days—4 working, one free; in 1932 they changed it to 5 weeks of 6 days. Such experiments were abandoned in 1940.

The Babylonians based their calendar on that of the Sumerians. From a surviving letter by King Hammurabi about 1700 BCE, we know that the Babylonian year, consisting of 354 days, contained 12 months that alternated between 30 and 29 days—but from time to time they changed one of the latter to 30 days, and every three years or so they threw in an entire extra month. They did not number their years in regular succession, however, until the reign of King Nabonassar in 747 BCE. By about 500 BCE they had figured out that 19 solar years equaled almost exactly 235 lunar months—every 19 years the sun and the moon were in phase again. This interval was reported by the Athenian astronomer Meton upon his return to Greece from a visit to Babylon, whereupon it was called the *Metonic cycle*. For centuries after adopting the Babylonian calendar with its 354-day years, the Greeks assigned each year a Golden Number, designating its occurrence in the Metonic cycle.

Egyptian civilization, by contrast, began as early as the fifth millennium BCE to base its year on the stars, in particular the Pleiades (the Seven Sisters) and Sirius (the Dog Star). Invisible for months because of its nearness to the blinding disk of the sun, Sirius makes a brilliant reappearance in the eastern sky just before dawn on the day that the mighty Nile, the world's longest river, starts to rise. The coincidence of the Nile's annual four-month-long life-giving inundation with the rising of the Dog Star induced the Egyptians to use it as the foundation of their calendar—a year consisting of 360 days, with 12 months of 30 days each. (The division of the circle into 360 degrees no doubt originates from this 360-day year.) It did not take the Egyptians very long to notice that a year consisting of 360 days did not work very well, and so they changed their calendar to a 365-day year, divided into twelve 30-day months and 5 extra feast days. They then observed that even a 365-day year was not fully satisfactory: Sirius rose one day late every four years, though they did not regard this as sufficient reason to change their system. Just as the Chinese did not number their years but used their dynasties instead, so the Egyptians relied on counting the occurrence of a year in the reign of a given Pharaoh. (The Greeks, from the third century BCE on, counted Olympiads, intervals of 4 years or alternately 49 and 50 Greek months, with a starting date at the restoration of the games in 776 BCE from their much earlier origins.) Nevertheless, the introduction of the stellar calendar for civil purposes can be traced back as far as 4236 (or possibly 4241) BCE in Egypt, making this the earliest fixed date in history.

Darius the Great introduced the Egyptian 365-day year into Persia, and it was known to some Greeks. Thales of Miletus and Herodotus, for example, had learned about it from their travels to the land of the Nile in the sixth and fifth centuries BCE. Still, the Greek states did not adopt it. Meanwhile, religious calendars went their own way. The Jewish calendar, with intercalations to keep in step with the seasons and to make certain that feasts like Passover would occur at the right time, had years of various lengths. According to the Talmud, these were determined by the court of justice on an ad hoc basis by observing the progress of the crops. For example, an official document of Rabbi Gamaliel II, issued about 100 CE, states: "We make known to you that the lambs are small and the birds are tender and the time of the corn harvest has not yet come, so that it seems right to me and my brothers to add to this year 30 days." The present lengths of the Jewish years are governed by a 19-year cycle: a 13th month of 30 days is intercalated in the 3rd, 6th, 8th, 11th, 14th, 17th, and 19th year, after which the cycle is repeated. By contrast, intercalation of any kind was expressly forbidden by Muhammad in the Koran, though it is a matter of dispute whether this stricture was actually meant to eliminate a system of intercalations that had existed at an earlier time in the Arab world or whether it was based on ignorance of the need. In any case, the Koran not only forbade it but explicitly ruled out any future change of this sanction.

Another way of dealing with the problem of reconciling the lunar and solar cycles was adopted by the Mayas, for whom time was part of their religion. Using both the lunar month and the solar year, they simply kept separate records of both, without a system of combining the two. Once in a while they would go back to reconcile the accounts, and the result was a very accurate, if not very practical, calendar.

Rome's first calendar, produced by King Romulus about 738 BCE, consisted of 10 months, beginning at the celebration of the vernal equinox (the start of spring) and ending about January 24, which added up to 304 days; the additional 61 days were not counted until the next king, Numa Pompilius, introduced a lunar cycle and added two months. Since this new year had a length of 355 days, causing it to quickly fall out of phase with the seasons, an extra month of 22 or 23 days was added at the end of every other year. This calendar remained in use for about 300 years until 450 BCE, when Appius Claudius shuffled it to insert the month of Februarius, which had been situated at the end, between Januarius and Martius. The new calendar was not published, however, but was kept secret by the priests, who constantly tinkered with it, adding and subtracting days or months at will, until it got quite out of sync with the seasons. Order was finally established in 46 BCE, when Julius Caesar, after consulting the Greek-Egyptian astronomer Sosigenes, devised what came to be known as the *Julian calendar*—a public document rather than a state secret. After a "year of confusion" that lasted 445 days to bring it back into synchrony with the sun, he followed the Egyptian model of a 365-day year, but in order to correct for the one-day error every four years that the Egyptians had chosen to ignore, he introduced the leap year, so that in the regular year five months of 30 days alternated with six months of 31 days, except for Februarius, which had 29 and every four years, 30. After his death, renewed tinkering with the rhythm of the leap year caused his calendar to fall out of step again until it was brought back into line in 8 BCE by Caesar Augustus, who also renamed the month previously called Sextilis after himself. In order to avoid the bad luck associated with even numbers, he gave it 31 days, taking the extra day from Februarius. The resulting calendar continued to be known as "Julian."

Which finally brings us to our modern calendar. What Julius Caesar did not know was that an additional error of about 1/300 of a day in the length of a year would eventually throw his calendar out of step with the sun. The solar year in fact consists of 365.2422 days rather than 365.25, which caused the vernal equinox, as calculated by the Julian calendar, to occur earlier and earlier as the centuries passed. By the sixteenth century it was falling in early March, and eventually it would have shifted to Christmas time if nothing had been done about it. The concomitant change in the date of Easter Sunday[2] induced the Council of Trent to authorize the Pope to alter the calendar once more. In compliance, Gregory XIII ordered the day of October 4, 1582, to be followed by October 15, and the years from then on to begin with January 1, rather than March 25 as in the Julian calendar, with the extra day in leap years omitted one time per century in three out of every four centuries. Thus, 2000 and 1600 were leap years, but in the years 1900, 1800, and 1700, which would have been leap years without Gregory's adjustment, the extra day was omitted. It will be omitted again in 2100.

The Gregorian calendar was adopted throughout the

Christian world, except in Russia and Britain and her colonies. England did not follow until 1752, when the day after Wednesday, September 2, became Thursday, September 14. (By that time the old calendar had accumulated an error of another day.) Finding the bright side of all this in his *Almanack,* Benjamin Franklin wrote, "Be not astonished, nor look with scorn, dear reader, at such a deduction of days, nor regret as for the loss of so much time, but take this for your consolation, that your expenses will appear lighter and your mind be more at ease. And what an indulgence is there, for those who love their pillow to lie down in Peace on the second of this month and not perhaps awake till the morning of the fourteenth." Which is why the birthday of George Washington, born on February 11 according to the old calendar, is today celebrated on February 22.

Russia held out until 1917; the October Revolution actually occurred on November 7, according to the new reckoning. However, the Orthodox Church to this day has not adopted the Gregorian calendar and celebrates the New Year on January 7, nor has the Jewish calendar been changed. The Muslim calendar is still strictly lunar, with 6 months of 30 days and 6 of 29, making up a total of 354 days for the year. As a result, the number of Gregorian years since the beginning of the Islamic era differs now by more than 40 from the number of Muslim years.

Are we then at the end of the calendar story, or might there be a later need for readjustment? Well, the Gregorian calendar is accurate to within 26 seconds per year, an error that will add up to a day in 3,323 years. I think that will do for now.

3

Early Clocks: Home-Made Beats

For units of time smaller than a day, no natural rhythms offer guidance. The Saxons divided the day into "tides," testimony of which remains in the "morningtide," "noontide," and "eventide" of poetry. But neither the Greeks, Romans, nor Chinese used a subdivision of the day until the Near East taught them about hours. From the religious or priestly point of view, the durations of the day, month, and year were dictated by the heavens and therefore by the gods; any further subdivision had to be man-made and therefore suspect.

Presumably basing their units on a long-lost custom of the Sumerians, who divided the whole day into 12 periods (and each of these into 30 parts), the Babylonians early on divided the daylight-time and night-time each into six units. But it was the Egyptians who bequeathed us the 24-hour day. Beginning in the middle of the fourth millennium BCE, they used a 12-hour subdivision of daylight and a 12-hour night; the further division of the hour into 60 minutes and the minute into 60 seconds was, again, a Babylonian innovation. The Romans of 500 BCE marked only the sunrise and sunset, which a crier announced from the steps of the Senate building. It was not until 159 BCE that the hours of daylight were signaled for the public.

At the beginning of the Christian era, the Romans divided the period of daylight into 5 "hours," to which Pope Sabinianus in 605 CE added two more hours and ordered the church bells rung to mark them. These seven canonical hours remained the dominant divisions of the day in Europe for centuries, and they still determine the times of the services in Roman Catholic and Anglican churches. The night, however, in many places continued to be separated, as it had long been, into four "watches" or "bells." Whereas in the lower latitudes of the Middle East the separate division of the times of darkness and daylight into a fixed number of parts did not result in widely varying units, in the northern parts of Europe these "hours" had very different durations, depending upon the season.

The earliest means of measuring the partitioning of the day was, of course, the shadow of the sun, almost always visible during daytime in Egypt and Mesopotamia. Three methods served to indicate the time of day. The first was based on the length of the shadow cast by an upright stick, later called a *gnomon* by the Greeks, or by a tall monument. The second was based on the direction of that shadow relative to its direction at noon, when it is at its shortest. And the third was based on the position of the shadow cast on a marked beam by a raised cross-bar (see Figure 4). The best-known monuments used for showing either the length or the direction of the shadow were the Egyptian obelisks, symbols of the sun god Ra that were first built about 2000 BCE and served both as calendars (by measuring the length

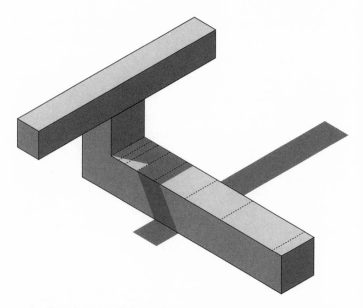

4. An Egyptian sun clock. In the morning, it was laid horizontally with the cross piece pointing toward the east; in the afternoon it was turned around to point toward the west.

of the shadow at noon) and as indicators of the hour. As gifts or plunder, New York, London, and Paris each have one obelisk, Rome several. The other shadow-based method employed the sun dial, which remained in use for thousands of years in Europe.

Since these time-measuring procedures were useless in the dark, another means had to be found for nighttime, and for many centuries the Egyptians told the time of night by the stars. Astral time-tables have been found in the tombs of Ramses VI and Ramses IX at Luxor, dating from the twelfth century BCE. The construction of such tables required another, independent, method for telling the time, and that was the water clock, which the Greeks later called the *clepsydra,* or "water-stealer." This timepiece consisted of a decorated water container that allowed its contents to drain through a small hole in the bottom, so that the level of the remainder, as measured by marks on the inner wall of the vessel, showed the elapsed time. The first known maker of a water clock was an Egyptian astronomer-physicist by the name of Amenemhet (the earliest astronomer known by name) in the middle of the sixteenth century BCE. Clepsydras were later used in the Roman Senate to cut off long-winded speakers. Refined and perfected in a variety of ways, they remained in use in Europe for some three thousand years.

The first improvement in the water clock was an attempt to correct for the fact that the rate of flow slowed down as the water level in the container got lower. A later version used an upper reservoir, from which water dripped into a funnel-shaped chamber kept filled at a constant level by means of an overflow; this chamber then gradually issued its contents at a constant rate into a third container at the bottom, whose rising level, in turn, raised a float indicating the time elapsed. A second enhancement was made in about 250 BCE by Archimedes, who attached the float to a gear arrangement that turned a pointer on what we would recognize as the face of a clock (see Figure 5).

In the course of centuries, the gear mechanism of the water clock was elaborated to include automatic corrections for the varying lengths of the hour as the season changed—remember that day and night each contained 12 hours, in both summer and winter. Decorative moving figures were added as well, to increase the clock's appeal (see Figure 6).

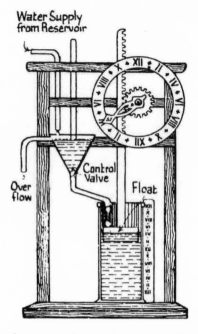

5. A clepsydra commonly used in Greek and Roman time.

Perhaps the fanciest of them all was reputed to have been presented by Harun al-Rashid to Charlemagne. In addition to artistic refinements, an important new element added to the water clock was sound, in the form of constant dripping, which foreshadowed the ticking of later mechanisms and, more loudly, town criers or bell ringers to announced the hours, who were subsequently replaced by automatic chiming, chirping, and ringing devices.

The water clock had an important drawback: it failed to work when the temperature was cold enough for water to

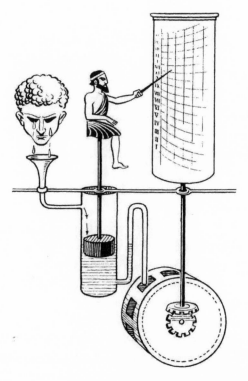

6. The clepsydra of Ctesibius, from about 250 BCE. As water dripped from the eyes in the head of a statue, a human figure rose, pointing at the hours indicated on a vertical cylinder. A syphon mechanism made both the figure descend to the bottom every 24 hours and a waterwheel turn, which then very slowly rotated the cylinder by 360° once a year. The markings on the cylinder adjusted the length of the hours according to the season.

freeze. A purely mechanical device was needed, especially in northern latitudes. Just as the basic principle governing the water clock was the constant rate of its flow under the influence of gravity, the mechanical clock that began to compete with it some time late in the thirteenth century used the steady pull that gravity exerts on a weight hanging freely from a cord wound around an axle. The resulting rotation was transmitted by a set of gears to a pointer on a dial and to a variety of more and more complicated mechanisms controlling bells or other audible, visible, or mobile devices. The trick was to make the central axle turn as evenly as possible, and the clever device designed for this purpose was the *escapement*.

The origins of this ingenious gadget are in dispute. For a long time it was credited to a Benedictine monk by the name of Gerbert, born about 920 CE, who possessed a vast, almost necromantic knowledge of astronomy and who would later become Pope Sylvester II. However, there also was the huge mechanical clock built about 1090 CE by the Chinese mandarin Su Sung, which incorporated as a central element the escapement mechanism that had been at the heart of a clock built in China some 300 years earlier.

Whether knowledge of this Chinese invention had reached Europe is unclear, and we have no evidence that Gerbert was aware of it. For all we know, the escapement was independently invented in Europe and in China within a time span of about 150 years. In some other details, however, the mechanical clocks of China and Europe differed greatly, quite apart from the fact that Su Sung's big astronomical clock tower was ultimately water-powered, whereas in Europe the escapement was employed in weight-driven timepieces (see Figure 7).

7. A pictorial reconstruction of the clocktower built by Su Sung and his collaborators in 1090 CE. The clockwork, rotating both an armillary sphere and a celestial globe, was driven by a waterwheel and fully enclosed within the tower. Puppet figures gave notice by sight and sound signals of the passing hours and quarters. (Original drawing by John Christiansen.)

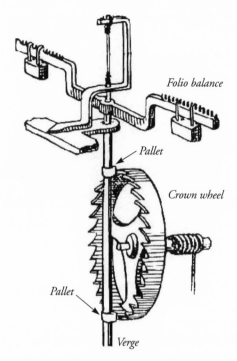

8. An early escapement with its crown wheel, verge, and folio. The asymmetrical shape of the teeth of the crown wheel allows the pallet to slide up on one side of a tooth easily, turning the wheel, and then to drop and engage the next.

In order to use the free fall of a weight as a driving mechanism for a uniformly running clock, it was necessary to circumvent the gravitational acceleration that the downward motion of the weight inevitably experiences unless constantly slowed. This is the crucial function of the escapement, whose principal parts are the *crown wheel* and the *verge* (see Figure 8). As the former is turned by the cord tied to the descending weight, it is engaged by a "pallet" attached on the side of the verge, which momentarily stops its motion. However, the torque exerted by the weighted cord on the crown wheel makes it begin to rotate again, and the one-sided slant on its teeth allows the pallet to escape, thereby rotating the verge, whereupon another pallet attached to the latter is caught in the crown's teeth, stopping it again. The constant repetition of this cycle thus prevents the crown wheel's rotation from accelerating. The slowly horizontally oscillating "folio" connected to the verge was an additional device to assure, by its inertia, the even turning of the crown wheel, whose fitful but overall unaccelerated rotation finally either turned the hand on a clock dial or drove a bell-striking or figure-turning mechanism, or both.

These mechanical clocks were so large and expensive to build and maintain that they were used initially only in churches and other public buildings (see Figure 9). In the Houses of Parliament at Westminster Hall in London, a predecessor of Big Ben was built at the time of Edward III (with repair records going back to 1381), and the great clock of Rouen in Normandy was built in 1389. When clocks eventually became sufficiently compact for domestic use, only the wealthiest houses could afford them. Though they became symbols of riches and power, these domestic clocks were wildly inaccurate timekeepers: in the course of a day they could be off by as much as half an hour or more and had to be reset.

But well into the seventeenth century, accuracy was not an important desideratum for mechanical clocks. Extremely elaborate decorations or ingenious mechanical devices

9. A sixteenth-century etching of a famous clock in Strasbourg.

driven by them were. Everyone was much more interested in the entertainment afforded by magically moving figures than in the accuracy of the time they told, and the enormous skills developed by artisans who built these ostentatious affairs became a tradition that well served the makers of more accurate timepieces, based on different principles, in succeeding centuries.

Figure 10 shows a modern reconstruction (following his detailed plans) of a "planetarium" or astronomical clock designed and built between 1348 and 1362 by Giovanni De'Dondi, a very inventive Italian physician and astronomer. Britten's *Old Clocks and Watches and Their Makers* describes this weight-propelled mechanical clock thus: "For complexity, it has seldom if ever been surpassed and the mechanical problems solved by De'Dondi, as far as one can tell out of his inner consciousness, must be regarded as nearly miraculous, when compared with the very scant mechanical achievements of his century."[1] Much of the gearing in this ingenious piece of apparatus has served as a model for other mechanical devices to this day.

Although the definition of the passage of time was based on the periodic motions of the rotation of the earth on its axis and its revolution around the sun, none of the early timekeeping and time-telling devices employed an intrinsically periodic physical phenomenon. Instead, they each used a process progressing at an even rate, ensuring by various means, as best they could, that this rate would indeed be constant—in the case of the clepsydra by attaching additional vessels to make the governing flow of water uniform, and in the case of the mechanical clock by adding the escapement. But without a periodic physical process, there

10. A modern reconstruction, at the Smithsonian Institution, of De'Dondi's planetarium.

was no way to keep the pace of any of these mechanisms, clever and inventive though they were, from wandering away from the sun's time and therefore, like our biological clocks, requiring constantly repeated entrainment. Whereas punctual timekeeping had earlier been of no great interest, with the upsurge of science after the Renaissance an accurate measurement of time was needed. The construction of truly stable clocks, however, was not possible before Galileo's discovery of the properties of the simple pendulum.

4

The Pendulum Clock:
The Beat of Nature

Born in 1564, two months before William Shakespeare, Galileo Galilei ushered in the scientific component of the great period we know as the Renaissance. The son of a mathematician and musician in Pisa, he grew up to be a pugnacious and acerbic man who readily made enemies, because of his unconventional philosophical positions and his feisty personality. In the science of mechanics he argued strenuously against the prevalent Aristotelian legacy, supporting his views by experiments that are replicated to this day by students in physics courses, with objects freely falling or rolling down inclined planes. The story of his dropping two cannon balls of unequal weight from the leaning tower of Pisa—to demonstrate to the philosophers that, contrary to Aristotle, objects of different mass fall with equal speeds—became a legend, though it probably never occurred. For seventeen years he flourished as a Professor of Mathematics at Padua, devoting most of his time to research on motion, but also elaborating on Archimedes concerning the buoyancy of submerged objects and, incidentally, inventing the thermometer. (Based on the expansion

Galileus Galileus   Florentinus

Superior   licentia
J 6   24.
                                          12
Eques Octauius Leoni   Roman° pictor fecit

11. Galileo Galilei at the age of 60.

of air as the temperature rises, this thermometer was not very accurate, but it nevertheless may be regarded as the first scientific measuring instrument.) In 1609, stimulated by the recent invention of the telescope in Holland, Galileo's interest shifted to astronomy. He immediately constructed a telescope on his own—much improved in power over the Dutch model—and was the first to use it to study the heavens.[1] He discovered, among other things, the largest moons of Jupiter and offered a detailed account of their motion as satellites circling that enormous planet. The publication of his astronomy book, *The Starry Messenger,* caused a sensation all over Europe and led to his appointment as mathematician and philosopher to the Grand Duke of Tuscany, as a result of which he changed the venue of his scientific work to Florence.

Though Galileo did not believe in the elliptical orbits of the planets advocated by his contemporary Johannes Kepler, his astronomical observations convinced him that the heliocentric system proposed by Nicolaus Copernicus, who had died twenty-one years before Galileo was born, was correct. He declared so publicly until the Church pronounced this system heretical and in 1616 instructed him to abandon his view. After Cardinal Barberini became Pope Urban VIII, Galileo nevertheless obtained permission to present an impartial description of the two systems—the Ptolemaic geocentric and the Copernican heliocentric—which led him to publish his book *Dialogue Concerning the Two Chief World Systems.* As he used his own observations to support Copernicus, this volume turned out to be hardly impartial, and it was immediately banned by the Church. In 1633 he was taken to Rome and ordered by the Inquisition to face trial for heresy (insubordination was the real offense). Thirty-three years earlier the philosopher and astronomer Giordano Bruno had been burned at the stake for heresy, though on quite different grounds, and the memory was still fresh in people's minds. To escape death, Galileo was forced to abjure any belief in a heliocentric system, though legend has it that he left the scene muttering "Eppur si muove" ("And yet it moves"). He was condemned to life imprisonment, a sentence that was immediately commuted to permanent supervised house arrest. Blind and still under house arrest, Galileo died in 1642, figuratively handing the torch of modern science to Isaac Newton, born later that same year. (Galileo's death was recorded according to the Gregorian calendar, whereas Newton's birth followed the Julian calendar, still in use in England at the time.) The Church's condemnation of Galileo's astronomical discoveries was finally declared an error by Pope John Paul II in 1992.

Galileo's first important scientific discovery was the property of the simple pendulum, ideally consisting of a heavy bob suspended by a light-weight cord: so long as it does not swing too widely, its period (the time it takes for each swing) is independent of the amplitude of its oscillation (the length of the arc of the swing). Contrary to the anecdote cited in the Introduction, Galileo probably came to his discovery via his interest in music, which led him to experiment with pendulums of varying lengths for their rhythms. His first application of the pendulum as a timing device was in medicine, to determine the pulse rate of ill patients by means of a *pulsilogium;* the pendulum was employed to time the pulse beat rather than the other way around, as the story had it.

Many years later, while discussing the curious amplitude-independence of the pendulum's period with his son Vincenzio and others interested in clockmaking, it occurred to Galileo that, when combined with an appropriate escapement, the pendulum would be an ideal timer at the heart of a clock, which they confirmed by experimental testing. Galileo himself employed a water clock for his famous experiments on motion; the pendulum clock was not developed until late in his life, when he himself designed an escapement for it, the details of which differed somewhat from the schematic one shown in Figure 13, below.

It was not until fourteen years after his death that a clock based on Galileo's plan was actually completed, and in 1667, on orders from Grand Duke Ferdinand II, Georg Lederle of Augsburg constructed a pendulum regulator of the great scientist's design and installed it in Florence in the large one-handed clock on the western façade of the tower of the Palazzo Vecchio, the former residence of the Medicis. There it remains to this day, still accurate to within one minute per week (see Figure 12).

Although the idea of using a pendulum as a timing device was not new—the Arab astronomer Ibn Yunis the Younger was reported to have employed it as early as the twelfth century, and there are sketches by Leonardo da Vinci showing the pendulum in such a role—only after Galileo's fundamental discovery of its fixed period of oscillation (its *isochronism*) did its distinct and special suitability become clear.

Figure 13 shows schematically how the combination of pendulum cum escapement works in the form used after 1670. The axle holding the escape wheel is driven by a cord or chain holding a weight, as it had been in earlier clocks (and still is in reproductions of traditional clocks)—in later times it would be driven by a coiled spring. However, its rate of rotation is regulated not by a verge and folio but by

12. The great one-handed clock on the tower of the Palazzo Vecchio in Florence.

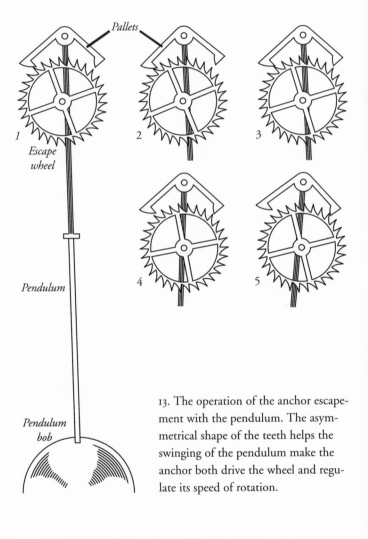

Pallets

1

Escape wheel

2

3

Pendulum

4

5

Pendulum bob

13. The operation of the anchor escapement with the pendulum. The asymmetrical shape of the teeth helps the swinging of the pendulum make the anchor both drive the wheel and regulate its speed of rotation.

pallets on an anchor rigidly attached to the pendulum, swinging freely at a constant frequency, thereby constraining the escape wheel to turn at the prescribed speed (producing the familiar tick-tock sound). At the same time, the torque exerted on the escape wheel by the weighted cord transmits a small kick to the pallet attached to the anchor every time it engages one of the teeth of the escape, thus preventing the pendulum from slowly being brought to rest by friction.

Experiments by the Dutch physicist Christiaan Huygens led to the discovery that the period of oscillation of the pendulum is proportional to the square root of its length; in order to double the period, the length has to be quadrupled. Thus, the period could be easily adjusted by raising or lowering the bob at the end of the pendulum. Huygens was thirteen years old when Galileo died. The son of a poet, composer, and diplomat serving the house of Orange, he was raised in a home that entertained cultural luminaries such as René Descartes. After studying both mathematics and law, and supported for many years by an allowance from his father, he devoted his life to scientific experiments and the study of nature. He made seminal contributions to the wave theory of light, and in astronomy his discoveries include the rings of Saturn and the planet's largest moon, Titan.

From a practical point of view, Huygens's most important contribution to the development of the pendulum clock consisted of an idea which improved its accuracy and reliability to such an extent that he, rather than Galileo, is sometimes credited altogether with its invention: he designed a clever cycloidal suspension, which automatically compensated for the fact that when the amplitude of the oscillation of an ordinary pendulum becomes too large its frequency changes somewhat. The best pendulum clocks later incorporated this device. Thus was born the first mechanical timepiece capable of being truly accurate. It became enormously influential during succeeding centuries. Because its beat, like that of the heavens, was governed by a strictly periodic physical motion, it required no repeated entrainment; when perfected, it was completely self-reliant and could be made very precise.

Today it may strike us as amusing that when pendulum clocks, in the form now known as grandfather clocks, became more widely available for the wealthier homes, people at first complained that they were inaccurate. Contrary to the earlier custom of dividing night and day each into an equal number of hours, no matter the season, these newfangled machines chimed more hours during winter nights than during summer nights! Prior to the Renaissance, people simply felt no need for an evenly running measure of time.

Indeed, from a modern perspective, one might well ask, what *defines* a uniform flow of time? The swinging of the pendulum provides this definition for us, and it was adopted because of its immense usefulness in the post-Renaissance world, both for scientific and practical purposes. The time-flow defined by the pendulum clock, in fact, as we now understand, differs somewhat from that which gave rise to our awareness of its rhythm in the first place, the apparent revolution of the sun around the earth and that of the earth on its axis. Why?

Because the earth circles about the sun, the time from

one sunrise to the next is not exactly equal to the period of its spinning about its axis. A thought experiment helps us understand how the motion of earth in its orbit, by itself, influences the length of day and night. Imagine that the earth did not spin at all: nevertheless, it would still be on one side of the sun at one time and on the opposite side half a year later; consequently, at any given location half the year would be night and the other half day. To get from night to day, the sun would have to rise once a year, and to get from day to night it would have to set once a year. So the earth's orbit alone has some influence on the time of sunrise and sunset, independent of its rotation on its axis. Furthermore, the precise difference between the length of a day and the spinning-period depends on the variable speed of the earth in the course of its elliptical (not circular) orbit around the sun. The velocity of the earth speeds up when it is closer to the sun and slows down when it is farther away (Kepler's second law tells us exactly how it varies). As a result, the time from one sunrise to the next also changes, and the time-flow defined by a sundial differs somewhat from the *mean time* defined by a uniformly swinging pendulum. The relation between the two is called the *equation of time* (see Figure 14). With this formula, one can translate the time shown by a pendulum clock into the time as indicated by a clock based on the sun at any time of the year, and vice versa.

Though in principle it would have been equally possible to adopt as the most fundamental definition the time-flow of the sundial rather than that of the pendulum, this would have been inconvenient and impractical in the extreme. Had it been done, Newton's equations of motion expressing

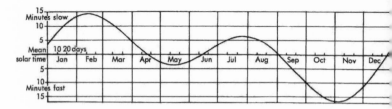

14. The equation of time. The mean time is slow as compared to the sundial time where the curve is above the central line, and fast where below.

his second law, which imply that a simple pendulum swings at a uniform rate, would have had to be different; the laws of physics would have led to a pendulum-period varying from season to season, and all subsequent development of physics would have been severely impeded. (We have here a good example of the interdependence of conventions and what are regarded as fundamental physical laws: though Newton's equations of motion are laws of nature, their validity is based on the adoption of a specific concept of the flow of time. For Newton, however, time was given by God.)

Because the Renaissance marks the beginning of truly world-circling trade and exploration by ocean-going ships, it is no surprise that the most pressing practical need for an accurate timing device came from navigation. A stable and reliable clock on board would enable a captain to ascertain the longitude of the position of his ship at sea. Its latitude—that is, its distance north or south of the equator—was relatively easy to find from the elevation of the sun at noon at a given time of year, or of the Pole Star at night.

But to determine how far east or west a ship was from its home port in London or Madrid or Venice was much more difficult. The method most commonly used was *dead reckoning,* which depended on measuring the direction of travel by means of a compass and estimating the speed of the ship from time to time by dropping a wooden log overboard and determining how quickly it fell behind, thereby keeping track of the ship's location by tracing its progress on a map.

A much more accurate way to ascertain a ship's longitude—which, together with its latitude, would fix its exact location—was first suggested in 1522 by the Flemish physician and astronomer Reinier Edelgestein, known as Gemma Frisius. He said that a stable clock should be carried on board which could reliably keep track of the time at the home port (or at any other fixed location of known longitude), rather than having to be wed to the local time shown by the sun or the stars. When the time shown by the clock was compared with local time as indicated, say, by a sundial, the difference would tell you by what fraction of 360° the earth had turned between having the sun directly overhead in the two locations, and therefore their difference in longitude. If you find, for example, that noon on board ship, as indicated by the sun, is an hour (= 1/24 of a day) late compared with noon in Lisbon (where the clock was originally set), you must be 15° west of there, because the earth must have turned by $360°/24 = 15°$ during the interval.

To keep accurate track of the home time, however, required a reliable shipboard clock that did not have to be constantly readjusted; repeated resetting, after all, would defeat its very purpose of staying faithful to the time in the ship's home port. Just as migrating birds, we now know, rely on an internal clock for navigation, so can our ships—though not for exactly the same reason. Whereas birds' circadian clock, together with the sun or the stars, helps them steer in the right direction, ships can use a compass to determine their direction of travel and a clock, together with the sun or stars, to fix their location.

Galileo specifically had in mind this urgent need for measuring the longitude of ships at sea when he suggested using the pendulum for regulating a clock (though he also proposed employing the observation of Jupiter's moons, which he had just discovered, for the purpose of measuring longitude), and so did Christiaan Huygens when working to improve that pendulum clock to achieve better accuracy.

How stable and reliable was this device? Could it be thrown off easily by small external perturbations? Idle in bed with a brief illness during the course of this work, Huygens noticed an odd phenomenon while watching the motion of two identical pendulums mounted in a single wooden case. Irrespective of how the two started out, after a while they always ended up swinging exactly 180° out of step—as one swung to the left, the other swung to the right. Not only that, but they became synchronous even if their individual periods differed slightly. (Recall that the internal clocks of biological systems are subject to similar synchronization effects.) While Huygens thought this sympathetic influence, evidently transmitted by tiny vibrations in the common support of the two pendulums (transmission through the air having been ruled out by experiments) might be employed to make a clock containing two pendulums more stable and accurate, it induced some members of the Royal Society, when they learned of it, to lose faith in ever being able to use pendulum clocks for the determination of longitude at sea. These timekeepers, they thought, were obviously much too sensitive to minute external influences.

An ordinary grandfather clock, of course, would not do for this purpose, anyway. The constant and sometimes violent motion on board ship, as well as the extreme variations in temperature and humidity during a voyage, would certainly prevent its mechanism from functioning reliably. The first thing needed, and not only for maritime purposes but also for more mundane uses at home, was a clock that could be put on a table or, even better, that could be carried around easily. The two components of the large pendulum clock that had to be changed to make it more compact and mobile were its two central elements: the weight propelling it and the pendulum regulating it.

The introduction of what is now called the *main spring* during the second half of the fifteenth century did away with weights. (The invention of the main spring is sometimes credited, with doubtful validity, to Peter Henlein of Nuremberg.) This coiled spring, made of elastic steel, was wound tightly around a drum with a key; its slow unwinding during the course of the day served the same purpose as the weight pulling on a cord. Because the main spring was invented before the pendulum-escape, clocks with a main spring required another mechanism—the *fusee*—to make them run evenly. Its details need not detain us because it remained in use for only a brief time. Though the early main-spring-driven clocks were much more compact than those propelled by weights, they were still wildly inaccurate, gaining or losing sometimes as much as an hour per day, and there was no way to make them reliable. The weight-drive was used in some grandfather clocks long after the general adoption of the mainspring (I remember my father, once a week, "winding" a tall pendulum clock that stood in our dining room, by raising its weight rather than by turning a key) and remains in use today in many reproductions, especially the still-popular cuckoo clock.

The mainspring invention made it possible to construct portable timepieces that could easily be carried around, even upside down. Such "clock watches," some more or less egg-shaped, were made first in Nuremberg in the sixteenth century. Sometimes highly decorated and worn proudly on a chain around a wealthy person's neck, they became a status symbol until the Calvinists and Puritans, objecting to such ostentatious display, began carrying them in their pockets. Thus was born the pocket watch, popular until well into the twentieth century.

Needless to say, these watches still required modifications and refinements, which they received from an abundance of highly skilled watchmakers in Switzerland. When Calvin became head of the Church in Geneva, he frowned on the display of all finery and forbade the production and wearing of jewelry. Consequently, many former goldsmiths and jewelers, facing unemployment, turned their attention to the production of watches. Thus Calvin became indirectly responsible for the subsequent long-lasting prominence of his home town's watch industry.

After Calvin's death in 1564, his austere commands were relaxed, and the local artisans began to combine their watchmaking and jewelry-fashioning skills to produce ever more luxuriously decorated timepieces. Artfully adorned and bejeweled watches became prized possessions of dukes and kings, who presented them to one another as favors on

15. A seventeenth-century watch, made by
the English watchmaker David Ramsey,
with a case made of Limoges enamel.

16. Abesses' watch made by Conrad Kreiser of Strasbourg, seventeenth century.

special occasions. Upon receipt of such a gift, they had their court watchmakers examine it, copy it, and if possible improve upon it. Thus the art of precision watchmaking was disseminated rapidly all over Europe, with London as one of its main hubs, but Geneva remained its most active center. Throughout the twentieth century, the symbol of an accurate and reliable timepiece was still a fine Swiss watch.

5

Successors: Ubiquitous Timekeeping

A portable replacement for the cumbersome long pendulum—which, like the propelling weight of a clock, functioned only in an upright position and could not be easily transported or carried about—was finally invented in the second half of the seventeenth century by Robert Hooke (a man we will meet again in conflict with Isaac Newton) and our old friend Christiaan Huygens. Their invention was the *coiled balance spring.* Its refined version, now sometimes called the *hairspring,* is a very thin steel spiral whose alternating tightening and unwinding makes a balance wheel attached to it rotate back and forth. Though the force it exerts comes from the elastic properties of a spring rather than from the earth's gravity as the pendulum's does, the underlying physics of this device is the same as that of the pendulum: they share the property of oscillating at a fixed frequency. The coiled balance spring's invention came none too soon.

In the foggy night of October 22, 1707, Admiral Sir Clowdisley Shovell and his officers on board the flagship *Association,* unable to pinpoint their location correctly and believing themselves to be west of Brittany, proceeded in a northeasterly direction in order to enter the English Channel. Instead, they hit the rocks known as the Bishop and Clerks off the Scilly Islands, west of the tip of Cornwall. The *Association,* along with three more of the Navy's finest warships, went down, drowning some 2,000 men, including the admiral (though he was later rumored to have been robbed and killed after being washed ashore alive). The disaster brought forcefully home to the Royal Navy the urgency of the need for a reliable method of finding the precise location of ships at sea.

The memory of the event and numerous petitions from English sea captains led Parliament, seven years later, to pass the Longitude Act, which promised a reward of £20,000 (a fortune at the time) to anyone who could devise a method for determining longitude to within half a degree. A degree of longitude near the equator is a distance of about sixty nautical miles, but this distance varies as the cosine of the degrees of latitude. In southern England, therefore, a degree of longitude is about 40 nautical miles; so the required accuracy—20 nautical miles—was not overwhelming. The Act established a Board of Longitude—whose membership included the President of the Royal Society, the Astronomer Royal, the First Lord of the Admiralty, the Speaker of the House of Commons, and the Savilian, Lucasian, and Plumian Professors of Mathematics at Oxford and Cambridge—to exercise complete discretion in awarding the prize and also to give monetary support to impecunious inventors with promising ideas. Proposals, of which there turned out to be a multitude, were to be tested on a voyage from Great Britain to the West Indies. For those longitude-measuring devices based on a clock, this meant that no more than 2 minutes could be gained or lost during the long voyage.

By the eighteenth century the distance corresponding to a degree of longitude was well understood and the oceans of the world and the outlines of the continents had been mapped to a certain extent. Some 250 years earlier, Columbus had made the gross error of estimating the distance from the west coast of Spain to the east coast of China to be only some 2,400 nautical miles, whereas we now know that the actual air distance is about four times as long. The reasons for his mistake were not only ignorance of an intervening continent but incorrect assumptions about the distance per degree of longitude and an overestimate of the size of Asia, which made its east coast appear closer. As a result of these errors, he set out on an adventurous voyage which he might not have undertaken had he known the true distance to his intended destination, a voyage which, of course, led to his stumbling on America. While he may not have been the first European to visit the western hemisphere, his "discovery" helped mapmakers much more accurately chart the surface of the earth.

In the eighteenth century, however, sailors were still handicapped by the lack of an accurate and reliable device for determining their location at sea, even though King Philip III of Spain had offered a prize of 10,000 ducats, Holland had put up 25,000 florins, and King Louis XIV of France 100,000 florins, all for a solution of the longitude problem. Galileo's proposal, which he had submitted in a letter to King Philip, had been rejected (along with a flood of crank ideas) as impractical. It was based on telling time by observing the moons of Jupiter, whose motion he had described in detail. Brilliant as the idea of using these moons like the hands of a fixed clock was, this heavenly dial was too difficult to see on board a heaving ship, where visibility was subject to the vicissitudes of the weather.

The British Longitude Prize was eventually won by John Harrison, son of a carpenter in Yorkshire, England, and a most inventive maker of wooden timepieces. At the time he entered the contest he had already built clocks that erred no more than a single second per month over many years. These clocks, of course, had not been subjected to the rigors of the sea. With some advice and a loan from George Graham, a famous London clockmaker who had been impressed by Harrison's preliminary design, he proceeded to construct what is now called H-1, his first candidate for entering the Longitude Prize competition. Six years later, it was ready to show to the Royal Society. Housed in a wooden cabinet and weighing 75 pounds, with gears made entirely of wood and four dials on a face decorated with carved cherubs, vines, and crowns, it was a smashing aesthetic success. When tested by the Board of Longitude in an abbreviated trial run, however, it did not measure up to the required accuracy. Still, it impressed the board sufficiently to persuade them to give Harrison £500 to make improvements—half of it on the spot and half upon completion of a new version. The venerable H-1, sans cabinet, is still running and on show at the National Maritime Museum in Greenwich.

In 1741 Harrison completed H-2. More modestly decorated, it was made of brass and placed in a smaller housing. Though much more resistant to temperature changes and other sea hazards, it was never officially tested, for two reasons. First, Harrison's presentation of it to the board was so self-deprecatory and critical of his own creation that it led members to believe he wanted nothing more than continuing support—which they supplied—for his efforts to further improve his timepiece. And second, England was at war with Spain, which put oceangoing vessels at risk. Consequently, Harrison started work on H-3, which took him nineteen years to complete, during which time he was supported by five grants of £500 each and was the honored recipient of the Royal Society's Copley Gold Medal (later awarded to Benjamin Franklin, Ernest Rutherford, and Albert Einstein), as well as the offer of a Fellowship in the Royal Society, which he declined. As the construction of H-3 progressed, Harrison's opinion of the size required for a clock of the needed reliability changed dramatically, and he began to conceive of a portable watch that might do the trick. He completed it in 1759, while the Seven Year's War was raging, and by the time the seas were calm enough for a trial run, he had completed a timepiece he much preferred; so H-3 was never submitted for a test.

H-4, the watch that John Harrison finally felt rightfully proud of, was only five inches in diameter and weighed three pounds. It was full of innovative touches, such as bearings made of diamonds and rubies to reduce friction, as well as decorative flourishes. To test the compact little marvel, Harrison's son William and two officials from the Board of Longitude sailed on the *HMS Deptford* from Portsmouth to Port Royal, Jamaica, leaving on November 18, 1761. To everyone's relief—the ship had run out of beer—she arrived in Madeira, the end of the first leg of the long

17. The dial and backplate of Harrison's watch H-4.

trip, well ahead of what the ship's captain had expected on the basis of dead reckoning but in precise accord with position calculations by means of the watch. When they arrived at Port Royal after 81 days at sea, a comparison between the local longitude as calculated on the basis of H-4 and as determined by astronomical means showed that the clock had lost only 5 seconds.

John Harrison should have won the Longitude Prize then and there. However, long drawn-out resistance by the Board prevented him from receiving more than a fraction of it. (The Astronomer Royal allegedly was competing for the prize himself and was said to have conspired against Harrison. His method of telling longitude relied on specially constructed lunar tables rather than on a mechanical clock.) It took another ten years, the intervention of King George III, and a special petition to Parliament before Harrison was awarded the full balance of the great prize at the age of eighty. Three years later he was dead.

The drawback of the winning entry for the Longitude Prize, as compared with its predecessors H-1, H-2, and H-3, was that its small size required a regular regimen of upkeep, such as oiling and periodic cleaning, in addition to daily winding with a key. (The others were engineered so precisely that they needed neither oil nor regular cleaning.) As a result, in contrast to the others now shown in the National Maritime Museum, H-4 is exhibited not as a running, functioning clock but at rest; over the years it would have suffered too much damage by these intrusions.

The design of the modern marine *chronometer*—a word reputed to have been coined by John Arnold, another watchmaker of prodigious productivity—is based on improved versions of John Harrison's path-breaking watch, later refined by the Swiss Ferdinand Berthoud, and another model constructed by Pierre Leroy of Paris. Marine clocks based on these designs remained in constant and universal use for some 250 years. But today the coordinates of any point on the surface of the earth can be ascertained instantly to within a few feet via satellites of the Global Positioning System (GPS). Though not based on a mechanical oscillator like the chronometer, it also functions by means of precisely synchronized clocks, one in each of three satellites and a fourth at the point P on the earth whose position is to be found. These are used for the measurement of the exact distance from P to each of the satellites by determining the time it takes for a radio signal to traverse the three distances; triangulation then allows the location of P, both its longitude and its latitude, to be calculated geometrically, by computer. For the system to work anywhere on earth required the installation of a large network of permanently stationed satellites, so that at least three of them are reachable by straight lines from every position. But this, of course, is jumping way ahead of our story.

To make it sturdy for constant carrying under conditions of everyday use, the modern mechanical watch, its size enormously reduced and its reliability much improved, incorporates two marginal refinements. The first is a balance wheel made of two metals that expand at different rates as the ambient temperature rises, or else made of special alloys less sensitive to temperature changes. The second is a further improvement on the crucial escapement (see Figure 18), the most delicate part of the mechanism and the one subject to the greatest wear because it produces about

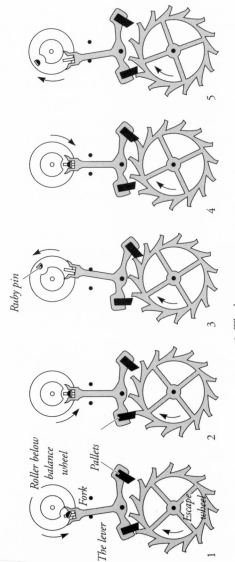

18. The lever escapement.

150,000 ticks per year. This modification facilitated the construction of extremely thin watches. In order to reduce friction and wear, the principal bearings are made of semi-precious stones; the number of "jewels" is usually taken as an indication of the quality of the watch.

The availability of reliable and accurate watches and public clocks became indispensable during the nineteenth century because of the development of the railroad. In earlier periods the demand for public and personal time-telling had arisen primarily from governmental functions, such as the sessions of the Senate of the Roman empire, and religious observances—calling the flock together for church services—but these needs did not require timepieces of great precision. In fact, many public clocks did not even have minute hands, as the great clock on Palazzo Vecchio in Florence still testifies. For the most part, only maritime navigation and scientific work—especially astronomy—required more precision than the hour of the day. It was the construction of networks of trains in Europe and America, with their interdependent schedules, that made watches accurate to within a minute indispensable. The dark-suited, visor-capped conductor holding his impressive pocket watch in front of him and calling for the train to start moving, not a minute late, became the symbol of punctuality. And if you did not want to miss that train, you had better have a reliable watch yourself.

The railroads necessitated not only precision but also a certain geographical uniformity in the telling of time. Clearly, schedules of a wide-ranging network of connecting trains set according to local times as determined by the sun on the spot were a great inconvenience; there had to be a general time agreed to by all. After years of confusion, an international conference held in 1884 in Washington, D.C., adopted the venerable Greenwich observatory near London, founded in 1675 by King Charles II, to be the location through which the zero meridian runs, so that all longitude measurements would begin there. Moreover, it was agreed to choose Greenwich Mean Time (GMT), the precisely determined local mean solar time at the Greenwich observatory—the local sidereal time modified by the equation of time—as the standard "home" time for navigation all over the world. This setting of the hands of the clock, also called *coordinated universal time,* would serve well for most of western Europe, since for this geographical area it did not significantly differ from the local time. (Actually, the clock time adopted for general convenience by most of Europe is one hour ahead of GMT.) On the North American continent, on the other hand, the east-west distances were so large that the sun set four hours later on the west coast (seven hours on the west coast of Alaska) than on the Canadian east coast, and a single standard time would not do. Since the rest of the world had been carved up into 24 different time zones, each 15° wide, the continent fell into eight time zones (four in the U.S. mainland), within each of which the conventionally agreed-upon local time was uniform, and universal train schedules could be set up. Even as late as the middle of the twentieth century, there were pockets of rebellious individualism, however. The clocks in Indianapolis varied by an hour from those in Bloomington, an hour away to the south by car, and residents of Bloomington had to be careful to take this difference into account when trying to meet a train or plane.

Such idiosyncrasies no longer exist, but after the almost universal adoption of daylight-saving time in the summer —first introduced in 1918 to conserve precious hours of daylight and save energy—the state of Indiana still refuses to go along, remaining on Eastern Standard Time all year round. The states of Arizona and Hawaii have also refused to adopt daylight-saving time.

An unavoidable consequence of the existence of different local times at different longitudes, or of the establishment of various time zones, was the appearance of a *date line*. When it is 10 P.M. on Tuesday in Nome, Alaska, it is 1 A.M. on Wednesday in Los Angeles (45° east and hence 3 hours ahead of Nome), 4 A.M. in New York, 10 A.M. in London, noon in Moscow, 7 P.M. in Beijing, and 9 P.M., Wednesday, on the Kamchatka Peninsula. So, although Beijing and Nome are only three hours apart according to their clocks, they differ by a day on their calendar. Travelers from Los Angeles to Beijing do not need to adjust their watches by more hours than they would when flying from London to New York, but they do have to change the date.

The existence of such a fissure in the calendar along some line running from the North Pole to the South Pole is inevitable, but its location is quite arbitrary. By international convention it is drawn along a meridian—more or less— that runs through a little inhabited region in the Pacific, where it causes the least inconvenience. For the local island population, however, the consequences are a bit weird. If you sail from Samoa to the Fiji Islands, leaving late Monday night, you will arrive early Wednesday morning, having skipped all of Tuesday; but if you set out Wednesday evening to go back to Samoa, you will return there early Wednesday morning, regaining the lost day.

The level of clock precision required for running rail-roads was insufficient for later scientific work as well as for everyday uses such as plane travel, and timepieces had to be developed to accommodate these needs. The most important change in modern clocks and watches, beginning in the middle of the nineteenth century, was the use of electro-magnetism in a variety of forms. First, an electric motor was added to do the work of winding, a time-consuming task in the case of large clocks. Before the installation of a large motor to perform this chore automatically, it took three men five hours to wind Big Ben in London, and they had to do it three times a week—45 man hours of work performed 52 times a year.

A more significant use of electricity came when clock-makers began to exploit the vibrations of the musical tuning fork to keep time. These vibrations are governed by the same physical principle as the oscillations of a spring or a pendulum; like the latter, the fork has its own proper frequency, which is much higher than that of any swinging bob of a grandfather clock. Producing an audible hum, it vibrates at a rate of several hundred cycles per second. The unit of vibration is the Hertz, named after Heinrich Hertz, a German who discovered radio waves in the second half of the nineteenth century; the proper frequency of a clock pendulum is about 1 Hz, while that of a tuning fork is in the range of 400 Hz. These vibrations can be mechanically transferred to the rotation of a wheel to turn the hands of a clock by a mechanism analogous to a pendulum escape (see Figure 19), while the humming of the fork is maintained by periodically administering electromagnetic kicks. This con-traption, commercially called the Accutron, is a timer of great accuracy.

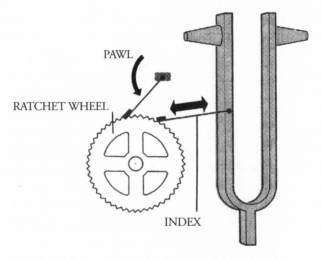

PAWL

RATCHET WHEEL

INDEX

19. A tuning-fork timer. The vibrating tine of the fork pushes a small spring, called the index, against a ratchet wheel that turns the hands of a clock. The pawl prevents the wheel from turning backwards.

Even better results were achieved by the quartz oscillator, based on a crystal. Nature keeps the faces of a crystal in place by elastic restoring forces similar to those acting at the ends of a steel spring or on a tuning fork. Consequently, just like Galileo's swinging lamp, the balance spring of a mechanical watch, and a tuning fork, a crystal has a proper frequency of oscillation irrespective of the amplitude of its vibration, except that the number of cycles per second in this case is very much higher even than that of a tuning fork, namely in the millions of Hertz (MHz). Quartz, a very common and abundant crystalline form of silica, the sand on dunes and deserts, is especially useful because it was found to be *piezoelectric:* that is, it generates an electric voltage when compressed, and a voltage in the opposite direction when stretched.

Suppose then that a flat quartz crystal is covered on both sides by metal plates to which an oscillating electric voltage is applied. The alternating electric forces will tend to compress and stretch the crystal, inducing it to vibrate reluctantly—"reluctantly," because of its internal restoring forces and the ever-present friction—at the same frequency as the applied field. The piezoelectric property of the crystal will, in turn, generate an oscillating electric field in response. If the applied voltage happens to alternate just right, at the proper frequency of the crystal, the crystal will shed its reluctance and vigorously vibrate on its own, producing a field that reinforces the one applied. The result is called *resonance* and the arrangement, a *resonator.*

The principle of using a quartz resonator at the heart of a clock is analogous to that of the pendulum with escapement. In the case of the pendulum, the torque originating from an axle—turned either by a weight-driven cord or a mainspring—transmits a small kick to the pendulum to keep it swinging, while the proper frequency of the pendulum regulates the rate of rotation of the axle. In the case of the quartz resonator, the applied voltage keeps the crystal's vibration going, and the proper frequency of the vibrating quartz regulates the oscillation of the electric field produced. The great advantages of the quartz oscillator over a mechanical pendulum are that the former can be made quite small as well as mobile, it requires little energy to keep it going, and it works with enormous accuracy. The quartz clock in the Greenwich observatory, which replaced the earlier mechanical one, is so stable that its total number of cycles per day—at millions per second—varies by no more than two from one day to the next.

Of course, the voltage oscillations of the resonator have to be translated into a visible time indicator. This is done either by purely electronic means, resulting in a numerical display on a liquid crystal or via light-emitting diodes, or else by having the generated voltage drive a small electric motor rotating in synchrony with the alternating voltage and turning the hands on the dial of a watch. Timepieces made in this way can be inexpensive, durable, and very accurate. In addition, two such clocks can be made very precisely synchronous, as is needed for the functioning of the global positioning system.

While the natural vibrations of a piezoelectric crystal are an accurate and relatively stable means of measuring time, they are not entirely reliable over long periods. As a crystal ages, the period of its vibrations decreases, throwing the clock off. The color of the light emitted by an atom—that is, the frequency of the electromagnetic waves making up the light—on the other hand, is absolutely fixed. This is the principle underlying an *atomic clock*. One form of such a device works as follows.

An atom of the element cesium (a metal similar to sodium) consists of a central nucleus surrounded by a cloud of 133 electrons, each of which resembles a little magnet. When the direction of the outermost electron-magnet is reversed, which can be done by means of an externally applied electromagnetic field, the energy of the atom rises by a certain fixed amount; after a short while that magnet will relax back to its normal direction, with the atom expelling the energy difference in the form of fluorescence, emitting a quantum of electromagnetic radiation of a frequency of exactly 9,192.631770 MHz, in the range of radio waves. To reverse the direction of the electron-magnet in the first place takes a quantum of radiation of that same frequency. So if an electromagnetic beam of radio frequency is directed at a bulb full of cesium gas, all of whose atoms have the magnets of their outermost electrons oriented in the normal direction of lower energy, and the dial changing the frequency of the beam is turned, at the precise point at which the magnets are yanked around, a large portion of the beam will suddenly be absorbed. Much of its energy will be taken up by reversing the magnetism of the atoms' outer electrons, and the absorption is readily detectable as a diminution of its intensity after emerging from the bulb. In this manner the radio frequency can be tuned precisely to the cesium "resonance," and the beam can be used to drive a clock at the rate determined by that frequency.

The only remaining inaccuracy is caused by the normal irregular motion of the cesium atoms in a gas at non-zero temperature. In order to overcome this, the latest version has six infrared laser beams gently pushing the cesium atoms together into a ball and propelling them upward, like a fountain, until the ball comes to rest and begins slowly to fall back under gravity. The radio-frequency beam is directed at the ball of cesium atoms just when it is at rest at the top of its trajectory. This is called a *fountain clock,* the most accurate form of the cesium clock in existence at this time. The pace at which time advances is thus fixed and determined, stably and absolutely, by a standard intrinsic to nature. This is how the unit of time—the duration of a second—is now internationally defined.[1]

We have come a long way from Galileo's legendary fascination with the swinging of a lamp in the cathedral of Pisa. But all the procedures for measuring time, so important to many aspects of modern life, both commercial and scientific, utilize at their core scientific principles identical to those governing Galileo's pendulum and its oscillations. When we look at those underlying physical laws, we shall find consequences that have profoundly modified not only our conception of the flow of time but also our notions of the very structure of the world.

In order to understand the basic physics of the properties of the pendulum that enabled it to make its revolutionary contributions, we have to change gears from a largely historical mode of description to a scientific one. We also have to go back again more than three centuries to the time immediately following Galileo's death.

6

Isaac Newton:
The Physics of the Pendulum

Though Galileo discovered the isochronism of the pendulum as a fact of nature, he did not offer an underlying reason for his seminal observation. That explanation had to wait for the great work of Isaac Newton.

Born after his father's death in Woolsthorpe, Lincolnshire, on Christmas day of 1642 (January 4, 1643, according to the Gregorian calendar not yet in use in England at the time), Isaac Newton was raised mostly in his grandmother's house. "A sober, silent, thinking lad" who grew up to be a solitary, lonely man of unpleasant disposition, he loved to tinker (building water clocks, among other things) and draw diagrams.[1] His family had been completely without learning, and his mother had tried to make him into a farmer. But after a grammar-school education taken up almost entirely with Bible studies and Latin, he was encouraged by an uncle to pursue his education further. In 1661 Newton managed to enter Trinity College, Cambridge, as a "subsizar," a poor student supporting himself by performing menial tasks for others. In 1665, the year he should have

20. Isaac Newton at the age of 83. (Portrait by Enoch Seeman.)

graduated with a Bachelor's degree, the university closed for almost two years because of the plague, and he had to return home for most of that time.

These were the *anni mirabiles,* during which Newton, working in total isolation, laid the foundations of most of his seminal work in mathematics, celestial mechanics, and optics. Upon his return to Cambridge, where he remained for 30 years, he was made a Fellow of Trinity College, and, at the age of 26, the Lucasian Professor of Mathematics. Though Newton corresponded with other leading scientists of the day, he continued to work feverishly, mostly in solitude, on theological speculations (becoming a fervent follower of the Arian heresy, a belief that could be very risky for his future in Cambridge), on experiments in alchemy, and on his magnum opus, *Philosophae Naturalis Principia Mathematica.* With financial support from his friend, the astronomer Edmund Halley, this great tome was published after considerable delay in 1687 and was instantly recognized throughout Europe as epochal. In it, Newton introduced the law of universal gravitation, the laws of motion, and the calculus, a mathematical procedure which he had invented especially for use in physics and which became the seed for a vast and enormously fruitful branch of mathematics. Together, the laws or motion and gravitation united the falling apple with the orbiting earth and the motion of the moon.

After the publication of the *Principia,* Newton began to make more contacts with the outside world. He was elected a member of Parliament from Cambridge University (both Cambridge University and Oxford University have their own representatives in Parliament), he met Christiaan Huygens, with whose wave theory of light he disagreed, and he quarreled almost constantly with other prominent scientists. He had a row with Robert Hooke, another disputatious member of the Royal Society, after Newton published results obtained by Hooke without giving proper credit, and he unjustly accused Gottfried Wilhelm Leibniz, who had independently invented the calculus, of plagiarism. The dispute contributed to a long-lasting alienation between British and Continental mathematicians; in the end, the mathematical world would adopt Leibniz's approach and notation rather than Newton's. The strain caused by Newton's work and his continual quarrels led him to suffer a severe depression, after which he began to pursue other interests. In 1699 he was appointed Master of the London Mint, and three years later he was elected President of the Royal Society, a position he retained until his death. Shortly thereafter Queen Anne knighted him, and he was now Sir Isaac.

In 1704 Newton collected the results of his researches on the properties of light by publishing *Opticks*. He had been the first to construct a reflecting telescope, and his experiments had led him to the important new conclusion that white light consists of a spectrum of many colors. However, he turned out to be wrong about the fundamental nature of light: he believed it to be made up of solid particles. Huygens's wave theory was closer to the truth, though it too had to be modified some 200 years later. Newton, founder of the modern Scientific Revolution, "standing on the shoulders of giants," in his own phrase, and continuing where Galileo had left off, died in 1727. He was given a state funeral and buried in Westminster Abbey, as Voltaire observed, like a king.

In order to understand the underlying reason for the property of the pendulum discovered by Galileo, we have to remind ourselves of Newton's laws of motion. Applied to this special case, they yield a complete description of the pendulum's oscillation, including the fact that its period does not vary with the width of its swing, so long as that width remains small.

Newton's new physics completed the final break with Aristotle that Galileo had begun. Whereas the ancient Greek philosopher had taught, and the world had believed him for some two thousand years, that objects would not move unless a force was applied, Newton's first law stated that when no force is acting on it, an object will remain either at rest or in a state of uniform, rectilinear motion; the effect of a force is to *accelerate* it. (In the wider sense of the word, as used in physics, "accelerate" may mean not just speeding up but also simply changing the direction of motion.) Specifically, his second law asserts that to produce the acceleration a, a force of magnitude $f = ma$ in the same direction as a is required, where m is the mass of the moving body. If there are several objects exerting forces on one another—Newton's third law demands that if object A exerts a force on object B, then B exerts a force of equal magnitude and opposite direction on A—his second law applies to each of them. To solve these equations so as to find the positions of all the bodies in the course of time, given their initial positions and initial velocities, requires the use of Newton's differential calculus. The next 250 years saw an explosion of new developments and discoveries based on his work, both in physics and mathematics.

Before actually using Newton's second law—the equation of motion—and applying it to the specific case of a swinging pendulum, we can deduce Huygens's observation that the period p of its motion is proportional to the square root of the length of the pendulum by the following reasoning, which physicists call *dimensional analysis*. The only basic quantities entering the problem are length, time, mass, and the strength of the force of gravity, which may be measured by the acceleration g of free fall. Recall that, according to Galileo, this acceleration is the same for all objects. Every freely falling body increases its speed every second by 32 ft/sec. Acceleration is a change of velocity per unit time, and speed is distance per unit time; therefore the "dimensions" of the acceleration g are (distance/time)/time $=$ distance/time^2. Since Galileo had discovered that the period of a pendulum is independent of the amplitude of its swing, the only relevant quantities available are the length of the pendulum l, the acceleration of gravity g, and the mass of the bob m, and the only way to obtain the period p—a quantity of the dimension "time"—from these is to divide l by g, getting a "time" squared because the distances cancel, and then, in order to obtain a "time," taking the square root of the result: p must be proportional to the square root of l/g. (Notice that the dimensional argument tells us also that the period of the pendulum cannot vary with the mass of the bob; there simply is no other relevant quantity with the dimension "mass" that could serve to cancel m so as to end up with a pure time.) Indeed, solving Newton's equation of motion in detail, as we shall do in a moment, leads to the conclusion that $p = 2\pi\sqrt{l/g}$. Neither of these arguments, however, was used by Huygens, who discovered his law by experimentation.

In order to explain the motion of a simple pendulum swinging by a small angle and to describe that motion in detail, we will have to apply Newton's second law and do a little mathematical work. (Any reader who is allergic to mathematics may skip the next couple of pages.) As Figure 21 shows, the force F points in the direction toward its equilibrium position and thus opposite to the displacement of the bob from the center. The similarity of the two triangles

21. Forces on a pendulum. The weight-force W produces both a tension S on the string and a horizontal force F dragging the bob back toward equilibrium. For a small swing, the actual path of the bob, which is a circular arc, is approximated by a straight line as in the figure, and b is approximately equal to l.

implies that $F/W = x/b$, which is approximately equal to x/l (as the figure shows, because x is much smaller than l) so that the magnitude of the force F tending to accelerate the bob equals $F = xW/l$. On the other hand, if the same bob were freely falling instead of being attached to a string, it would experience the acceleration g and Newton's second law would require that $W = mg$, which implies that the magnitude of the horizontal force on the pendulum bob equals $F = xm(g/l)$.

Thus the equation determining the motion of a simple pendulum, or *harmonic oscillator,* says that its acceleration is proportional to the distance from the center and in the opposite direction. The farther the bob swings, the more it is slowed down, until the motion stops and turns around. Since from then on the acceleration is in the direction of the motion, it speeds up until the bob gets past the equilibrium position, where the displacement becomes negative and the acceleration positive, again slowing it to a stop and turning it back. Moreover, the magnitude of its acceleration at the distance x from the center is $a = F/m = x(g/l)$.

The solution of this equation can be represented pictorially (see Figure 22). As the bob swings with varying speed, the swing distance x, plotted against the time, is a multiple of the oscillating sinusoidal curve shown in the figure and repeated over and over to the left and right. This curve is a sine function: $x = A \sin y$. The constant A, which is the amplitude of the oscillation of the pendulum—its maximum swing—is not determined by the equation but by the initial velocity when it is set in motion or the angle at which it is let go. On the other hand, y is proportional to the time t, and the speed with which y increases determines the period

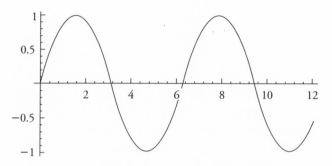

22. A plot of the function sin y.

p of the pendulum, that is, the time it takes for the pendulum to perform a full swing, which equals the time for the angle y to complete a circuit and increase by $360° = 2\pi$ in radians. (The measure of y in radians is the length of the crust of a slice of pizza of opening angle y and radius 1, which means $360°$ corresponds to 2π.) It therefore follows that $y = 2\pi(t/p)$, so that $y = 2\pi$ when the time is equal to one period. Instead of using the period of the oscillation, one may also characterize it by its *frequency*, the number of beats per second. If it does two swings per second, it obviously takes half a second for the pendulum to perform a full swing; if it does three swings per second, a full swing takes one third of a second: the frequency f is the reciprocal of the period p, $f = 1/p$. As a result we can write $y = 2\pi ft$.

An important mathematical property of the sine function $x = A \sin(2\pi ft)$ is that its acceleration at the time t is given by $a = (2\pi f)^2 A \sin(2\pi ft) = (2\pi f)^2 x$. From the result stated above, namely, $a = F/m = x(g/l)$, it therefore follows that $(2\pi f)^2 = g/l$, or $p = 2\pi\sqrt{l/g}$, a result in accord with what we obtained earlier by dimensional analysis. (This ends the mathematical interlude, and allergic readers can relax.)

In addition to explaining the detailed behavior of a harmonic oscillator, the Newtonian equation of motion, embodying his second law, implies another important property of a swinging pendulum, which the French physicist Jean Bernard Léon Foucault discovered by keen observations on the pendulum-driven mechanism for rotating the camera (the daguerreotype had just been invented) that he used for his astronomical work. Foucault was born in 1819, the son of a bookseller and publisher. A small, frail man who lived in Paris all his life, he made several important scientific contributions, one of which was the first laboratory measurement of the velocity of light, both in vacuum and in water.

The property of the pendulum he discovered was that the plane of its motion—the vertical plane containing its fixed suspension and the line along which its bob moves—appears to turn slowly about the vertical line through its suspension point. Recognizing that the origin of this apparent rotation was the fact that this plane was fixed in space as required by the "conservation of angular momentum" implied by Newton's laws of motion, he concluded that it demonstrated in the most direct and visible way the rotation of the earth about its axis. That this is so can be most easily grasped if one imagines a pendulum suspended over the North Pole; the earth simply rotates on its axis under it, but because the pendulum's plane is fixed in space, to an observer standing on the earth it is the plane, not the earth, that appears to turn. The details of this effect are a little harder to visualize for other locations, but the principle is the same. Suspending an enormous pendulum from the dome of the Panthéon in Paris, Foucault arranged for a spectacular public demonstration of this phenomenon. Today, there are large Foucault pendulums exhibited in science museums around the world. The one in the Panthéon is still there, too. This discovery, and its understanding on the basis of the conservation of angular momentum, led Foucault to the invention of the gyroscope, a device that turned out to have extremely important applications for navigation. He died in 1868 at the early age of 48.

The harmonic oscillator—the abstract version of Galileo's swinging chandelier—and the sine function of Figure 22 describing its motion made an appearance in almost all areas of physics after Newton. Historically, the origins of this function lie in geometry, with particular applications to astronomy going back to the third century BCE in Greece. Its use in dynamics, for the description of motion and the behavior of physical systems in time, was a modern innovation with wide repercussions that began with the analysis of the pendulum by means of Newton's second law of motion. As far as classical mechanics is concerned, the principal reason for the ubiquitous appearance of this oscillator function is that for an attracting force to vary in proportion to the distance from a center, that is, linearly with x, is the very simplest situation possible. Such linear, or proportional, behavior applies equally well to the force exerted by a stretched spring and a compressed crystal—which is why the balance spring could eventually replace the pendulum to produce a portable, reliable mechanical watch, and the quartz crystal could, in turn, replace the balance spring, utilizing electricity.

What is more, even if an attracting force on an object does not vary in proportion with the distance, it can in most instances be closely approximated, for small displacements, by a linear dependence. This is, of course, exactly what happens in the case of a pendulum: when it swings widely, the force pulling the bob back toward equilibrium is not simply proportional to the distance; the approximations—replacing the arc of the bob's swing by a straight line and replacing b in Figure 21 by l—are then no longer valid. Galileo's observation that the period of the oscillation is independent of its amplitude holds only for small amplitudes, when the angle away from equilibrium remains small during its entire motion. For larger amplitudes, the pendulum has to be modified in the clever way invented by Huygens, which assures that the restoring force remains linear even then.

The sinusoidal dependence on the time, therefore, is a very widely observed phenomenon describing almost all small oscillations found in nature. Their important characteristic—to repeat, because we will make use of it later on—is that at every instant the acceleration is equal to the negative of a multiple of the displacement, and that multiple is proportional to the square of the frequency, namely $(2\pi f)^2$.

In order to understand the even wider applicability of the harmonic oscillator, we have to look at a development in mathematics that took place almost a century after Newton's death. The mathematician primarily responsible for this was Jean Baptiste Joseph Fourier. Born in 1768 in the town of Auxerre, France, the son of a tailor, Fourier was orphaned at the age of 9 and educated in a military academy, where his interest in mathematics was first aroused. He grew up a many-faceted man with a strong appetite for politics and a talent for administration. During the French Revolution he was arrested by Robespierre for defending victims of the Terror, released, rearrested after the execution of Robespierre, and jailed for a short time for allegedly supporting him.

At that point Fourier fell under the spell of Napoleon. After joining the Egyptian campaign, he was appointed by the emperor to various diplomatic posts and high-level administrative positions. Though he was made a baron and eventually a count, Fourier resigned his posts in protest against the new regime of his benefactor after Napoleon returned from Elba. Thereafter, he devoted all his time to mathematical research, which he had previously been able to pursue only in spare moments. A member of the Académie Française and a Foreign Member of the Royal Society, he died in 1830 from the after-effects of a disease he had contracted in Egypt.

Fourier's remarkable discovery was that any arbitrarily given function of time can be expressed as a sum of sine functions of different frequencies. In other words, the sinusoidal behavior depicted in Figure 22 can be regarded as underlying any other kind of behavior with time in the sense that the latter is the result of adding up many such sine curves, all with different frequencies and amplitudes. The given function is completely characterized by its "Fourier coefficients," that is, the strength with which each frequency is represented in it.

For example, suppose an electric current is turned on instantaneously for one second to its full strength $-C$ and then reversed instantaneously for another second to $+C$. It

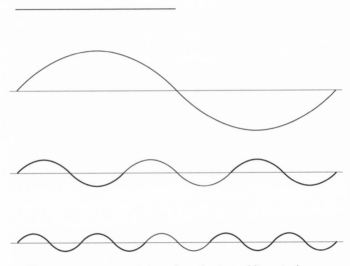

23. The upper curve, consisting of two horizontal lines, is the sum of infinitely many sine curves (each multiplied by a positive or negative constant), of which those shown are the first three.

can then be represented as a sum of sinusoidally oscillating contributions with the frequencies 1/2, 3/2, 5/2, . . . cycles/sec, in which these oscillations have the relative magnitudes 1, 1/3, 1/5, . . . , respectively, as shown in Figure 23.

Physically speaking, such a decomposition can be interpreted by visualizing any given time-varying phenomenon as being produced by a large—generally infinite—number of harmonic oscillators of different periods. Furthermore, in order to analyze it, in many cases it is sufficient, and much simpler, to confine the analysis to one single frequency at a time, for each of which the deceleration is proportional to the displacement multiplied by the square of the frequency. This simplification, turning many of their phenomena metaphorically into collections of pendulums, works most powerfully in two new fields of physics developed during the nineteenth century: acoustics and electromagnetics, to which we now turn.

7

Sound and Light:
Oscillations Everywhere

The protoscience of sound began with Pythagoras of Samos, the ancient Greek philosopher, mathematician, and mystic. He lived from ca. 560 to ca. 480 BCE, but his life and teachings are known only from mutually contradictory sources written some two hundred years after his death. After traveling widely in Egypt and Babylonia, he settled in Croton in southern Italy, founding a philosophical and religious society of considerable influence.

Pythagoras believed that "all things are numbers." Experimenting on the strings of his lyre, he developed the theory that the most beautiful musical harmonies correspond to the simplest ratios of numbers, such as 2:1 for the octave, 3:2 for the fifth, 4:3 for the fourth, and so on. This was the origin of what we know today as musical intervals and of the proto-science of acoustics, later contributors to which were Aristotle (fourth century BCE), the Roman engineer Vitruvius (first century BCE), and the Roman philosopher Boethius in the sixth century CE.

It was Galileo, however, who, about two millennia after

Pythagoras, transformed acoustics into a real science. He studied vibrations and described the relation between frequency of vibration and pitch, a relation that Boethius had already suggested. Galileo's discovery of the isochronism of the pendulum was also the result of experiments based on his interest in music; thus, the intimate connections between the study of sound and the properties of harmonic oscillations have deep, continuing historical roots. More detailed studies of the vibrations of stretched strings by the French mathematician Marin Mersenne, 1588—1648, on which his book *Harmonicorum Libri,* published in 1636, was based, became the foundation of what eventually became the burgeoning field of musical acoustics. However, a real understanding of the motions of vibrating strings had to wait until Isaac Newton formulated his laws of motion. The first to apply these to vibrations generally, forty years later, was the Dutch-Swiss mathematical physicist Daniel Bernoulli (1700–1782). The Bernoullis may be regarded as the Bach family of science: Bernoulli's father, his uncle, two brothers, a cousin, and two nephews all made important contributions to physics and mathematics, in addition to his own.

The wider science of acoustics came to full flowering with the work of the English physicist John W. Strutt. Born in 1842 at Langford Grove, Essex, Strutt was the son of a baron, whose title he inherited upon his father's death; Lord Rayleigh is the name under which he became widely known. Devoting his life entirely to the pursuit of science—it was extremely unusual at the time for a hereditary peer to engage in a profession outside the military, the government, or the church—he set up a laboratory in his family home. Rayleigh was elected President of the Royal Society and chosen Chancellor of Cambridge University; his work covered all of what is now known as classical physics, and some of it helped usher in the revolution in physics that occurred at the beginning of the twentieth century. Among many other publications, he produced a monumental two-volume treatise, *The Theory of Sound,* which remained the bible of the field for about a century. He died in 1919.

What is sound? What is the underlying scientific reason for the magical numbers introduced by Pythagoras? And what does sound have to do with the pendulum? Until the seventeenth century, the dominant view was that you heard the ringing of a bell because a stream of invisible particles emanating from the source reached your ears. This notion was definitively discredited by a famous experiment first performed by the German scholar Athanasius Kircher, described in his book *Musurgia Universalis,* published in 1650 and duplicated in innumerable lecture demonstrations ever since. Encasing a bell in a tightly closed jar and gradually removing the enclosed air with a pump, he had his audience listen as the ringing bell sounded dimmer and dimmer, until it eventually became inaudible. Lacking a sufficiently powerful pump to achieve a good vacuum, however, Kircher was unable to extinguish the sound completely and concluded incorrectly that air was not needed for the transmission of sound. It took the much improved vacuum pump invented by the Irish physicist Robert Boyle to perfect the demonstration ten years later. During the eighteenth and nineteenth century it gradually became clear that sound is, in fact, a wave in the air, consisting of minute changes in the ambient pressure and produced by vibra-

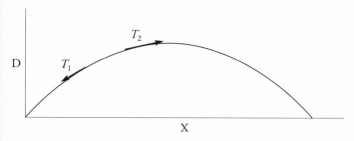

24. The forces on a piece of stretched string. (The curvature of the string is exaggerated in the picture.)

tions in church bells, explosions, vocal cords, musical instruments, and other sources.

To understand the physics underlying the production of sound by string instruments, let's look at what happens when a tightly stretched string is plucked (see Figure 24). (Again, a warning to the mathematically challenged: the next couple of pages will require a small amount of effort.) Given that each short segment of mass m is subject to Newton's equation $F = ma$, consider first the left-hand side. Pointing toward the equilibrium position, the force on any section of the bent string is the result of the fact that the two tension forces T_1 and T_2, pulling in almost opposite directions, are not exactly aligned—because the string is not straight—and thus do not cancel one another out. (The effect of gravity here is negligibly small.) This resulting force—as the effect of a variation in slope, with the slope being a variation in the displacement from equilibrium—is therefore proportional to the change in the rate of change of the displacement per unit length of the string, just as the acceleration is a change in its rate of change per unit time;

you might call it a pseudo-acceleration. The resulting equation, which has the pseudo-acceleration with respect to the distance x on the left and the real acceleration on the right, is called the *wave equation;* it occurs in many other contexts in physics.

At this point we avail ourselves of Fourier and fix our attention on a given frequency f, which allows us to replace the acceleration on the right-hand side of the wave equation by a negative constant times f^2, multiplied by the displacement D, whereupon it takes the same form as the equation of motion for the harmonic oscillator, but with the pseudo-acceleration as a function of x rather than the real acceleration as a function of the time t. Its solution is again a sine function and the displacement D of the string must be a multiple of sin (bx) (if the string is held fixed at $x = 0$ so that $D = 0$ there), with the pseudo-acceleration proportional to $-b^2$ times the displacement. Since the right-hand side is proportional to $-f^2$ and the left-hand side to $-b^2$, it follows that f is proportional to b.

There is, however, a constraint on the string of length L: its two ends are held fixed. This means that $D = 0$ not only for $x = 0$ but also for $x = L$. As the function sin (bx) vanishes for $bx = 0, \pi, 2\pi, 3\pi, \ldots$, it follows that the constant b is not arbitrary but has to have one of the values $b = \pi/L$, $2\pi/L, 3\pi/L, \ldots$, so that sin $(bx) = 0$ for $x = L$. (Figure 25 shows the shapes of the corresponding string vibrations.) Since we had found that f is proportional to b, the various frequencies, each proportional to the corresponding value of b at which the string can vibrate, also have to stand in ratios of whole numbers, as Pythagoras had correctly surmised. The first is called the *fundamental,* the second the *first harmonic,* and so on.

25. The fundamental vibration of a stretched string and its first and second harmonics.

The principles underlying the vibrations of the mouth-pieces of wind instruments and of the air columns in organ pipes are the same as for strings. For two-dimensional structures, such as percussion instruments, the situation is somewhat different. A metal plate, or the membrane of a drum, can vibrate in complicated ways with intricate nodal lines criss-crossing the surface where it remains at rest. The German lawyer Ernst Chladni (1756–1827) was the first to make these beautiful weblike patterns visible by sprinkling sand on a horizontal metal plate and stroking its edge with a violin bow—the sand collects where the plate remains still and is bounced off where it vibrates (see Figure 26). Napoleon Bonaparte was so fascinated by these demonstrations that he established an award for the first mathematician who could explain the figures, a prize won in 1816 by Sophie Germain, who, as a woman, had not been allowed to study at a university. In these cases, the *spectrum* of vibrations, that is, the allowed frequencies at which such two-dimensional surfaces can vibrate and set up the criss-crossing nodal lines, is much more complicated than for one-dimensional strings and depends entirely on the shape of the boundary of the plate or drum.

The strings on a modern musical instrument—similar arguments apply to the mouthpieces and the air columns in wind instruments and the surfaces of percussion pieces—

26. Chladni figures showing modes of vibrations of square plates.

transmit their vibrations to the surrounding air via *sounding boards* of much larger surface area; these produce small pressure variations—compressions and rarefactions—of the same frequency, which travel in the form of waves, spreading and eventually reaching our ears. Each pure musical note corresponds to a single frequency of a harmonic oscillation, vibrating somewhere between 20 and 20,000 Hz, or possibly including some *overtones*—harmonics—depending on the instrument. What is more, every sound, whether part of a cacophony or a melody, can be decomposed into its Fourier components; it consists of the superposition of harmonic oscillations of the air, each with a fixed frequency, some of them audible to us, others not—because their frequency is either too low or too high to be registered by our ears—as though underlying it all was a collection of Galileo's pendulums.

The other science initiated in the nineteenth century that extensively employs harmonic oscillations is electromagnetism. Knowledge of the existence of electric and magnetic forces goes back to Greek antiquity (*electron* was the Greek word for amber, the only substance then known to exert electric forces, and *e lithos magnetis* was the Greek term for the form of iron ore now called *magnetite,* which is found magnetized in its natural state in the ground), but such fragmentary knowledge did not become a serious branch of science until the middle of the eighteenth century (recall Benjamin Franklin with his kite in a thunderstorm), and our story does not begin until the nineteenth.

Born in 1791 in Newington, Surrey, the third son of a poor blacksmith, Michael Faraday (Figure 27) received little education—almost none in mathematics—and was sent to London at the age of 14 to be apprenticed to a bookbinder. A voracious reader, he devoured the works of Lavoisier (the French scientist usually regarded as the founder of modern chemistry), learned about electricity from an article in the *Encyclopedia Britannica,* and spent what little money he had to set up experiments on his own. At nineteen he began a

27. Michael Faraday as a young man.

more serious study of science at the City Philosophical Society and attended lectures and demonstrations by Humphry Davy at the Royal Institution, taking meticulous notes. At twenty-one, his bookbinding apprenticeship at an end, he had the good fortune to be asked by Davy, temporarily blinded by an accident in his laboratory, to help him out.

Young Faraday impressed the great chemist so much that he made him his permanent assistant and took him along on a two-year tour of France and Italy to visit the leading scientists of the day—a tour on which Faraday quickly gained an immense amount of scientific knowledge. Remaining at the Royal Institution for the next twenty years, where he made most of his ground-breaking discoveries in chemistry and electricity, Faraday grew into a famous and extremely popular public lecturer on science, drawing large audiences. At the age of 47 he suffered a nervous breakdown and did not return to scientific research for six years. The succeeding period of renewed productivity lasted ten years, after which his mind began to fail, possibly the aftereffect of poisoning during his earlier experiments in chemistry. For the last five years of his life, Faraday retired to Hampton Court, where he lived in an apartment provided by Queen Victoria. He died in 1867.

Though he was the quintessential experimenter, Michael Faraday introduced a seminal theoretical concept whose influence has been enormously fruitful in physics to this day and which will connect him to our theme of oscillators: the notion of a *field*.

When Newton formulated his revolutionary law of universal gravitation, he considered his own creation repugnant, as did everyone else. How could the distant sun exert a force on the earth, or the earth on the moon, without touching it? Such "action at a distance" repelled the mind. Faraday faced the same problem when studying the forces exerted by electric charges and magnetic poles upon one another, and the effect on them by electric currents. His novel solution was to replace the direct influence of one object upon another at a distant location by postulating a "condition of space" created by the first, propagating its effects by contiguous action from one point to another and exerting a force on the second object directly at the position of the latter. What he envisioned resembled rubber bands, which he called *field lines,* whose reality was, of course, not to be taken literally. In one form or another, the concept of electric and magnetic fields, as well as of other kinds of fields, including gravity, as conditions of space is still the fundamental idea underlying almost all areas of physics. What Faraday's version for electric and magnetic forces lacked was a mathematical formulation that would develop its full power, and this was provided by a Scottish physicist who was in his twenties when the great experimenter began to fade.

James Clerk Maxwell, son of a lawyer, was born in 1831 in Edinburgh and grew up in his parents' country seat, Glenlair, in the Maxwell estate of Middlebie. A great science enthusiast, John Clerk Maxwell strongly encouraged his son in the same direction, and after the death of his wife sent the boy to be educated in Edinburgh, first at the Edinburgh Academy (at the age of 15 he had his first mathematical paper read at the Royal Society) and then at the University of Edinburgh. After graduation from Cambridge University and a stint as Professor of Natural Philosophy at Marischal College, Aberdeen, James Maxwell was appointed Professor of Natural Philosophy and Astronomy at King's College in London. The death of his father brought him back to his family home in Scotland, where he remained for six years, pursuing his research. At the age of forty he returned to Cambridge, becoming the first Profes-

28. James Clerk Maxwell.

sor of Experimental Physics, and founded the Cavendish Laboratory. In 1879, he died of cancer at the age of 49.

Maxwell's profound contributions to physics covered several disciplines, including astronomy and the kinetic theory of gases, but the areas of most interest for our story, where he found a new kind of oscillation of the Galilean kind, are electricity and magnetism. Originally quite separate, these two fields of science were now known to be intimately connected. Contributions to this understanding had been made by three scientists: Michael Faraday, who demonstrated that a moving magnet produces an electric current in a wire; the French physicist André Marie Ampère (1775–1836), who found that current-carrying wires exert magnetic forces upon each other; and the Danish physicist Hans Christian Oersted (1777–1851), who discovered that a wire carrying an electric current exerts a force on a magnet. Maxwell was especially intrigued by Faraday's field lines, which emanated from electric charges and magnetic poles and surrounded electric currents. Taking many years to complete the task, he formulated a set of equations that fully characterized the field Faraday had been inspired to introduce and thereby unified the two disparate areas of electricity and magnetism into one: *electromagnetism.*

Before seriously attacking the problems of electricity and magnetism, Maxwell had been fascinated by color vision and the properties of light. (In 1861 he produced the first color photograph.) By this time, light was no longer generally regarded as a stream of particles, as Newton had believed, but was recognized as a wave phenomenon, as Christiaan Huygens had first proposed. There had been two decisive tests. The first was the discovery, in 1801, of the phenomenon of interference by the English physicist Thomas Young (1773–1829).

When two wave trains are superposed, they will interfere with each other in the sense that they may add up or subtract from one another, as shown in Figure 29, depending

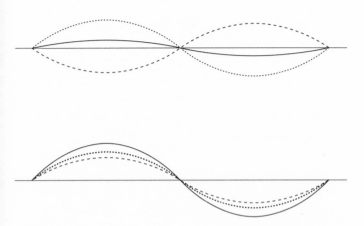

29. Interference between two waves of equal wave length. The lower figure shows constructive interference, the upper, destructive. In both figures, the curve drawn in a solid line is the sum of the other two.

on their relative phases. (Recall the notion of *phase* defined in connection with biological clocks in Chapter 1. In the lower figure, the two waves are exactly in phase; in the upper figure, they are 180° out of phase.) What Young had done was to pass light of a single color emerging from a small hole through two closely spaced narrow slits. The resulting image he observed on a screen consisted not only of two bright projections of the slits but a pattern of light and dark lines, which he interpreted as *interference fringes* produced by the alternating constructive and destructive interference of the light emerging from the two sources. Their relative phases depended upon the distances from these slits to a given point on the screen: when two wave trains, in phase at the slits, meet on the screen, the phase of the wave that traveled a longer distance will be behind that of the other, and what shows on the screen is their superposition. No stream of little corpuscles would give rise to such an image. Young's demonstration has served ever since as the most convincing demonstration of the wave nature of light.

The second experiment was a measurement of the speed of light in water, as compared with its speed in air. Newton's particulate theory predicted that light should propagate faster in water than in air, whereas wave theories led to the expectation that it would move more slowly. Measurements of the speed of light had previously been performed only by astronomical means, but in 1850 both Foucault and the French physicist Armand Hippolyte Louis Fizeau (1819–1896) separately measured the speed of light in water and in air with great ingenuity in their laboratories, and their results disagreed with Newton: light moved more slowly in water. (This speed in air or a vacuum is about 300,000 km/sec; that is why it is so hard to measure.) While these experiments tended to confirm the wave theory, their persuasive power did not equal Young's interference fringes.

So by the time Maxwell, fascinated by the properties of light, turned his full attention to electricity and magnetism, light was well recognized as consisting of waves, and the speed c with which these waves moved in a vacuum had been measured with great precision. But waves of what? Sound was understood to be an undulatory phenomenon too—compressions and rarefactions of the air. As light required no known medium for its existence, a medium was invented and named the *ether:* a massless, unobservable, ubiquitous, and all-penetrating substance in which waves propagated and made up what we call light. (*Ether* had been

Aristotle's name for his fifth element, which filled the space above the earth.) In contrast with sound, light waves were known to be *transverse;* like the wave on a vibrating string, they swing in a plane at right angles to their direction of motion (whereas sound waves are *longitudinal*). What is more, Faraday had discovered that the direction of this plane, called the *polarization* of the light, could be rotated by a magnetic field, indicating that light had a close relationship to magnetism. In addition, the number c (the speed of light in a vacuum) appeared as a numerical factor in Ampère's law and also in Faraday's law (both of which connect the electric current in a wire loop to the magnetic field flux through the loop). Maxwell therefore expected to find that light is intimately connected to the electric and magnetic fields whose physical properties he was trying to describe mathematically.

On the basis of an elaborate mechanical model of the structure of the ether, he was finally able to construct a set of differential equations that completely characterized both these fields together—henceforth called the *electromagnetic field*—and published his results in a *Treatise on Electricity And Magnetism* in 1873. The mechanical model he had used as a scaffold to erect his structure was soon discarded as superfluous, but his equations endured as an experimentally well-confirmed theory. Which takes us back to the subject of this book.

Subjecting the Maxwell equations for free space to the kind of analysis introduced by Fourier, one finds that for any given frequency f of harmonic oscillation, there are sinusoidal solutions of the wavelength $L = c/f$. In other words, the equations predict the existence of harmonic os-

GALILEO'S PENDULUM

cillations—ubiquitous Galilean pendulums, so to speak—
in the all-pervading ether. Those with frequencies in the
range of about 0.5 to 0.7 × 10⁹ MHz (0.5 to 0.7 billion
MHz), or wavelengths between 400 nm and 700 nm (one
nanometer equals one billionth of a meter), are visible to
our eyes as light, whose color depends on its wavelength or
frequency. Waves with lower frequencies, or wavelengths
in the range of centimeters or meters—experimentally discovered in 1890 by the German physicist Heinrich Hertz
(1857–1894), after whom the unit of frequency was named
—are now known as radio waves; those of much shorter
wavelengths, discovered later, are called x-rays and gamma
rays.

As a result of Maxwell's great work, we now have time
standards based on the frequencies of harmonically oscillating electromagnetic waves, which enable these standards to
be easily communicated from one place to another. If, after
landing on Mars, an astronaut's electronic watch malfunc-
tions, its rate—the length of its seconds—can be reset from
earth by simply sending an electromagnetic signal of known
frequency. Moreover, it can be synchronized with Greenwich Mean Time—so that it reads 12 o'clock noon when it's
noon at the Greenwich observatory—again by an electromagnetic signal. (Since the time light takes to travel from
one planet to the other is not negligible—light requires
about 8 minutes to reach earth from the sun—this has to be
done in two stages: first, a light signal is sent from earth
and, upon arrival on Mars, is instantly returned by re-
flection; the delay between sending and rearrival at home
allows us to infer the one-way travel time. When the signal
sent from Greenwich at noon arrives on Mars, the time

shown by a clock there should then be noon plus the light's travel time.)

There is, however, one big fly in the ointment: the postulated ether. Even though Maxwell's elaborate model for it had been discarded, it was still, in some way, assumed to be present everywhere. In a famous experiment, the American physicists Albert A. Michelson (1852–1931) and Edward W. Morley (1838–1923) tried to find this ether by determining how fast the earth is traveling through it. Their idea was analogous to the way one might determine how fast a submerged submarine is moving in the water: by sending a sound signal outside from one end of the vessel of length L to the other and inferring its speed s from the travel time T. Since sound propagates in the sea at a fixed known speed S, a signal from stern to bow should take the time $T_1 = L(S + s)$, and a signal from bow to stern should take the time $T_2 = L(S - s)$; therefore, s is easily calculated from $T_1 - T_2 = 2Ls$. In a similar manner, Michelson and Morley tried to determine our speed in the ether, through which light moves with the fixed known speed c, by carefully (presumably the earth moves very slowly compared to light) measuring the speed of light in various directions (in contrast to the sub, we do not know our direction of travel with respect to the ether) to see by how much it varied. To their great disappointment and astonishment, they found no difference at all. What was going on?

The puzzle was solved by the young German physicist Albert Einstein, whose solution had revolutionary implications for the measurement of time and hence for the ticking of a clock. Born in Ulm, Germany, in 1879, the son of a less-than-successful merchant, Einstein attended school in

30. Einstein in the patent office in Bern.

Munich after his family moved to that city. Chafing under rigid discipline, he did not do particularly well in class, but after some delay he was able to enter the Eidgenössische Technische Hochschule (Institute of Technology) in Zurich, graduating in 1900. After spending a year as a high school teacher, he landed a position as patent examiner at

the Swiss Patent Office in Bern (which had the added bene-
fit of allowing him to become a Swiss citizen). He also entered a marriage that, 16 years later, would end badly.

Though able to pursue physics only in his spare time,
Einstein managed to publish, during his *annus mirabilis*
of 1905, three articles with startling and seminal implications. One was an explanation of the photoelectric effect,
which had been discovered eight years earlier by Hertz and
whose strange properties, found by the German physicist
Philipp Lenard (1862–1947), had defied all attempts at ex-
plication; this paper contained the seeds of the *quantum
theory*. The second contained an explanation of *Brownian
motion,* which made the erratic, microscopically visible
movements of tiny dust particles in a liquid the first con-
cretely observable evidence for the molecular constitution
of matter; this article earned him a Ph.D. degree from the
University of Zurich (the ETH did not, at that time, award
doctoral degrees). The third of Einstein's papers of 1905
contained the *special theory of relativity,* with its completely
novel implications for the nature of time and space.

Understood and appreciated by very few people at first,
these publications nevertheless led to the offer of a junior
position at the University of Zurich in 1909, a full profes-
sorship at the University of Prague in 1911, and finally to an
appointment as the Director of the Institute of Physics at
the Kaiser Wilhelm Institute in Berlin in 1914. In 1915 Einstein published his theory of gravitation, called the *general
theory of relativity,* which differed from Newton's and predicted the bending of starlight by the sun's gravity. When
the size of this effect, calculated by means of Einstein's theory, was confirmed during a solar eclipse in 1919—observed

by the British astronomer Arthur Eddington during an expedition to Africa undertaken for this very purpose and initiated while the First World War was still raging—Einstein became world famous. Reviled by German Nazis as a Jew during the 1920s, he was celebrated during his travels abroad. After Hitler ascended to power in 1933, he refused to return to Berlin and accepted a permanent position at the newly founded Institute for Advanced Study in Princeton (especially set up in large part in order to accommodate him), where he remained until his death in 1955.

The central assumption of the special theory of relativity is that the speed of light in a vacuum is the same for all observers. Though it embodies the negative outcome of the Michelson-Morley experiment, this postulate originated, according to Einstein's later recollection, in a long thought process that began with his imagining, as a youth, what the world would look like if he were to ride on the back of a light wave. It clearly required a radical rethinking of the nature of space and time: if, moving with the speed s almost equal to the speed c of light, you followed a light signal, how could you still see that signal moving with the same speed c?

To solve this riddle, Einstein purged his thinking of all preconceptions and began from rock-bottom assumptions: in every laboratory where physical measurements are performed—physicists call these *reference frames*—time is defined by clocks and distance by yard sticks. An idealized reference frame is assumed to have a grid of marked distances and synchronized atomic clocks laid out so that the time and location of every event can be reliably fixed. (The atomic clocks are assured to run at the same rate, and they

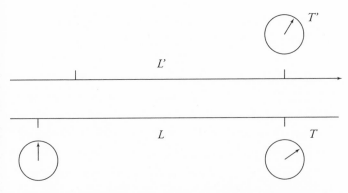

31. The lower figure shows the clocks of a stationary observer at the start and end of a light signal. The upper figure shows the clock of the moving observer at the time of arrival of the same signal.

can be made synchronous by means of light signals sent from one to another.) The bases of time and distance measurements in two reference frames that are in motion with respect to each other are therefore quite independent—there is no *God's time;* time is what clocks measure.

Having thus set the stage, we are ready to compare the results of measurements in two different laboratories. Figure 31 shows the clocks of a stationary observer at time 0 when a light signal is sent to the right and at time T when it arrives at a point a distance L away. At the start, the clock of the second observer, moving with the velocity s, also shows the time 0, but when the signal arrives, the clock there shows the time T'. In the meantime, the point at which the signal started has moved, so that the distance it traversed is L', less than L. Intuitively we would say that T' must be equal to T—all the clocks of the moving observer are synchronous with one another and with those of the stationary observer. And since from $L' = L - Ts$ it follows that the speed of the signal as seen by the traveler is $c' = L'/T$, we obtain $c' = (L/T) - s = c - s$. Einstein, however, insists that $c' = c$—the speed of light is the same for both observers—which is possible only if $T' = L'/c$, implying that T' *is less than* T. (The notion of an ether as the medium which light causes to vibrate and in which it moves at the velocity c fell victim to Einstein's incisive scalpel and had to be abandoned.) In other words, as seen by the stationary observer, *the traveling clock is slow.* Not only that, but the farther to the right you go, the more a string of traveling clocks, set up to be synchronous by the traveler, appear to differ from the stationary clocks they pass (because the larger L is, the more L' differs from L). This, of course, implies that the traveling clocks are *not synchronous* from the stationary point of view. Two events regarded as occurring simultaneously in one system of reference do not happen simultaneously when recorded in a moving system.

Note, by the way, that the situation is completely symmetric between the two observers: a clock carried by the locomotive of a moving train falls more and more behind as it passes a string of synchronous clocks arranged along the track, and a clock at a station falls farther and farther behind the passing clocks set up in each car and synchronized by the traveling train conductor. Therefore, the conductor has to conclude that the station clock is slow, while the station master concludes that the locomotive clock is slow. (The engineer's explanation of why his clock keeps falling behind the track clocks is that those are not synchronous, and the station master accounts for the fact that his clock seems to be falling behind the train clocks by saying that the clocks on the train as not properly synchronized.)

If you find these consequences of the theory of relativity startling, you are not alone; to this day, the *Journal of Mathematical Physics,* of which I am the editor, occasionally receives submissions of papers for publication whose authors claim to prove that Einstein has to be all wrong. The implications and predictions of this theory, however, have been abundantly confirmed with the greatest accuracy by many experiments during the course of the twentieth century. Why, you may ask, had no one ever noticed these strange effects before? The simple reason is that they are extremely small and difficult to measure so long as the traveling speeds involved are much smaller than the velocity of light, which at 300,000 km/sec is huge compared with all our everyday speeds. Under ordinary circumstances, the relativistic effects are much too small to be noticeable or even measurable, except by devices of enormous precision.

So where does this leave our nice, stable oscillator clocks? For Isaac Newton, space and time were absolute entities, ordained by God, and according to his mechanics, the progress of time defined by the simple pendulum, or by other oscillators based on the same principle, is universal and equally valid for all observers. According to the theory of relativity, however, two cesium clocks in motion relative to one another will each be seen by the other to be running slow. Indeed, if one of them is flown around the earth, it will, upon landing, be behind the one left on the ground—the experiment has actually been carried out on a globe-circling plane and the result confirms the prediction. This is also known as the "twin paradox": the twin returning from a long high-speed space voyage is younger than the one left behind. The symmetry between the two is here broken by the fact that the flying clock was necessarily accelerated. Whereas they serve to count the ticking of time at a universal rate at any two locations in the universe, even the best oscillator time-pieces cannot be made permanently synchronous if they are in relative motion. (The general theory of relativity implies, in addition, that a strong gravitational field also influences their rate of progress.)

If some of Albert Einstein's ideas led to radical changes in the relation between harmonic vibrations and the flow of time, other profound insights of his eventually engendered the understanding of an entirely new role for such oscillators—they constitute the very basis of matter. Up to this point, Galileo's pendulum has served only as a regulator of time, though with many important ramifications. As we shall see in the next chapter, it will now play an even more fundamental role in the universe.

8

The Quantum:
Oscillators Make Particles

In order to understand the fundamentally new role played by Galileo's harmonic oscillators in the physics of the twentieth century, we have to make a little detour.

The photoelectric effect, discovered in 1897 by Heinrich Hertz, quietly initiated a revolution in physics as well as many important technological applications. Hertz demonstrated that when light falls on a metal surface, it emits electrons; if the metal is connected to an electric circuit, these electrons cause a current to flow. Five years later, Philipp Lenard found that this photoelectric effect had a surprising property: whereas the number of electrons emitted increased with the *brightness* of the light, their speed depended only on its *color* (the shorter the wavelength, the faster the electrons move). Intuition and classical physics had led to the expectation that a more intense illumination would be able to eject faster particles, not just more of them.

No one was able to account for these experimental facts until 1905—the very year in which Lenard received the Nobel Prize for their discovery—when Einstein published his paper proposing their explanation in terms that would shake the foundations of physics. Some twenty years later, when the after-effects of Einstein's explanatory paper were beginning to unfold, Lenard, now an ardent Nazi, would publish violent polemics in favor of "German physics" and against the "Jewish physics" of Einstein.

Einstein's explanation of Lenard's findings was based on a reinterpretation of a strange idea used in a different context. Max Planck (1858–1947), another German physicist, had been able to account for the way the frequency distribution of the electromagnetic radiation emitted by a "black body" changes when it is heated, but not without making a radical assumption about the behavior of light when emitted and absorbed by matter. (A black body is any entity that absorbs all of the radiant energy falling on it rather than reflecting it. The stars are examples of such light-emitting hot objects, and so are glowing pieces of iron immersed in a fire; their colors change with the temperature.) He explained, for example, the color of a glowing piece of iron immersed in a fire or the color of a star in the sky by making a radical assumption about the behavior of light when emitted and absorbed by matter. He could not explain the experimental data unless he assumed, contrary to the entire body of classical theory, that the energies of the vibrating molecules which emitted and absorbed light were restricted to discrete values and that the frequencies of light were proportional to these values. The constant of proportionality later came to be called *Planck's constant,* usually denoted by *h.*

Capitalizing on this new idea, Einstein proposed an account of the photoelectric effect on the basis of a more far-reaching interpretation than Planck had ever intended:

light of frequency f always comes in the form of *quanta* of energy hf, and not just at the point of its emission or absorption, as Planck had assumed. Contrary to the view— well accepted since Young's interference experiments—that light consists of waves, and without denying the persuasiveness of Young's fringes, Einstein boldly reintroduced a new version of Newton's long-discredited corpuscular theory. Light, it seems, had the strange property of consisting *simultaneously* of both waves and particles. The particles eventually came to be called *photons.* If such a hybrid seems counter-intuitive, it is no more so than Einstein's other idea of the slowing down of moving clocks, though the consequences of the his theory of the nature of light were much more profound even than those of the special theory of relativity. In fact, Einstein himself regarded his paper on the photoelectric effect as the only one of his 1905 publications that was truly revolutionary.

The discovery of many other experimental facts that could not be explained on the basis of classical physics led to an intermediate theory, which cobbled together disparate classical pieces with ill-fitting quantum notions. Finally, in the 1920s, a fundamentally new paradigm emerged, which came to be called *quantum mechanics.* Supported and encouraged by the Danish scientist Niels Bohr (1885–1962), who had already made his name with his own contributions to the auxiliary, temporary quantum theory, its principal architects were three physicists: the German Werner Heisenberg, the Austrian Erwin Schrödinger, and the Englishman Paul A. M. Dirac. Though the grand old man, Einstein, was not happy with the philosophical implications of this audacious creation by the young Turks, his objections were

32. Erwin Schrödinger, about 1946.

overruled by the phenomenal agreements found between the new theory and a large and rapidly growing body of experimental results. There was no arguing with overwhelming success.

Schrödinger, born in 1887 in Vienna, son of a manufac-

33. Niels Bohr, Werner Heisenberg, and Paul Dirac in 1962.

turer, was educated at the gymnasium, where he particularly enjoyed ancient languages, mathematics, and physics. He received his Ph.D. in physics and at the University of Vienna. During World War I he served in the Austrian army as an artillery officer, and afterward moved to Germany and, in 1921, to Switzerland, where he became Professor of Physics in Zurich. On the basis of his success with the new "wave mechanics," he was appointed Professor of Theoretical Physics in Berlin in 1927, succeeding Planck, where he remained until Hitler's rise to power in 1933. The next three years were spent as a fellow of Magdalen College at Oxford, after which he returned to Austria, taking a position in Graz until the German annexation of Austria in 1938. At that point, he left Austria to become a senior professor at the Institute for Advanced Studies in Dublin, Ireland. In 1956, now in poor health, Schrödinger returned to the University of Vienna, where he was warmly celebrated by his hometown. He died five years later, in 1961.

Werner Heisenberg, fourteen years Schrödinger's junior, was born in 1901 in Würzburg, Germany, the younger son of a professor of philology. Displaying outstanding talents in mathematics at school (and as a pianist), he studied theoretical physics at the University of Munich, receiving his doctoral degree in 1923. From 1924 to 1926 he worked with Niels Bohr in Copenhagen and then was offered the Chair of Theoretical Physics at the University of Leipzig. He remained there until moving to the Kaiser Wilhelm Institute in Berlin in 1941 to direct the ultimately unsuccessful German effort to develop nuclear energy and, possibly, a nuclear bomb; his exact role in the latter endeavor remains a matter of controversy. After World War II, Heisenberg was appointed Director of the Max Planck Institute for Physics in Göttingen, which in 1958 was moved to Munich. He retained this position until 1970 and died in Munich in 1976.

The third and youngest of the three principal architects of our most fundamental physical theory, Paul Adrien Maurice Dirac, was born in 1902 in Bristol, the son of a British mother and a Swiss father who taught French. At Bristol University, he studied mathematics and philosophy but took a degree in electrical engineering. In 1923 he transferred to Cambridge University, where he completed his Ph.D. thesis in 1926. After extensively traveling on the continent and meeting all the leading physicists of the day, he was elected a Fellow at St. John's College in Cambridge in 1927 and appointed Lucasian Professor of Mathematics at Cambridge University in 1932. A man of few words but deep physical insight, he moved to the United States in 1971, where he became Professor of Physics at Florida State University. Dirac died in 1984.

The grand new theory called quantum mechanics was Dirac's synthesis of two apparently altogether different theories introduced almost simultaneously by Schrödinger and Heisenberg in 1925, which were called wave mechanics and matrix mechanics, respectively, but which turned out to be mathematically equivalent. A large number of features make quantum mechanics radically different from previously accepted physical notions, both with respect to underlying philosophy and experimental predictions. The most relevant of these, for our purposes, is its treatment of the energy of a physical system.

When a gas is heated to a high temperature, as in the sun, its atoms emit light and other electromagnetic radiation, and each chemical element emits light of a well-defined color. When sent through a prism, as Newton had done, all the component frequencies are separated into a collection of *spectral lines* characteristic of every element, like a fingerprint. This is how we know the chemical composition of the sun and distant stars.

For the lightest chemical element, hydrogen, the atoms of which were known to consist of only a central nucleus and one much lighter electron, the various frequencies making up its spectrum had been meticulously measured and found to obey simple rules, easy to state but hard to explain. Bohr's great contribution had been to propose a planetary atomic model, with the nucleus as the sun and the electrons as satellites that moved in specific orbits with specified energies that followed certain rational arithmetic laws. At high temperatures, when atoms collided at great speeds, their electrons were kicked up to orbits of higher energy—the atoms were "excited"—and subsequently descended to levels of lower energy, according to Bohr. As the electrons returned to their normal orbit, an atom would shed its excess energy E by ejecting a photon of that energy. In Planck's relation, the atom would emit radiation of the corresponding frequency $f = E/h$. The observed rules for the emitted frequencies were, then, simply expressions of the laws governing the allowed energy levels of the electrons in an atom (and Planck's curious quanta, the result of the fact that each light emission originates from a single jump of an electron from one energy level to another). When the atom is in its ground state, that is, its lowest energy level, it is absolutely stable and will emit no radiation.

Very ingenious, but by all the rules of classical physics quite absurd. First of all, why are only these orbits permitted? Second, Maxwell's electrodynamics decreed that an electrically charged particle such as an electron, orbiting around a nucleus, had to emit electromagnetic radiation, thereby continuously losing energy and consequently spiraling inward toward the center. An atom could thus last for only a short period of time before it expired. Ignoring such objections, of which he was fully aware, Bohr simply laid down his rules ad hoc, and they were remarkably successful in "explaining" both a number of complicated spectral observations and the stability of atoms in their normal state. If you had asked why these rules should be obeyed, however, no answer would have been forthcoming, and Bohr provided no coherent theory on the basis of which they could be understood.

The new quantum mechanics of Schrödinger, Heisenberg, and Dirac provided just such a general theory, and it led precisely to Bohr's rules for the special case of atoms. Any physical system described by classical physics as having a continuous range of energies, when "quantized" (that is, subjected to the general laws of quantum mechanics), would be constrained to exist at a certain specific, calculable spectrum of energies and no others. For different systems, this spectrum may consist of a continuous range of energies or a set of discrete values or both. Moreover, two systems of the same kind—say, two hydrogen atoms—at the same energy (and with the same quantum numbers belonging to other variables) are totally indistinguishable. There can be no scratches or other markings that would permit us to say "This atom here is the same one that we saw earlier over there." A helium atom in its ground state is a helium atom in its ground state is a helium atom in its ground state. Nothing else can be said about it.

The new theory lent itself to an enormous number of verifiable predictions and was able to account for many data that had previously not been understood; it passed every test with flying colors. Einstein's philosophical misgivings concerning the nature of the "reality" described by quantum mechanics fell by the wayside under the weight of such successes.

How, then, does this apply to the harmonic oscillator? If the Newtonian equation of motion of a simple pendulum is "quantized," that is, subjected to the new rules, it turns out that its entire energy spectrum consists of an infinite number of discrete points. (The reason why you do not notice this when observing a grandfather clock or even an extremely precise quartz clock is that the spacing between the allowed energies is minute.) What is more, it turns out that these allowed energies form a ladder whose steps are all of the same size. That the levels are all equidistant is a most unusual feature, which, as we shall see, has physical implications reaching far beyond the behavior of swinging lamps and marine chronometers.

Shortly after 1925, the German physicist Pascual Jordan (1902–1980), as well as Paul Dirac, began to think about the fate of field theories under the new quantum regime, with particular attention to Maxwell's equations. The road to quantizing these equations for the electromagnetic field was facilitated by first subjecting them to the analysis originated by Fourier (as discussed in Chapter 7), after which the equations could be regarded as describing a physical system with infinitely many components, one for each frequency. Just as Maxwell had found, every such component then acts like a harmonic oscillator: that is how his equations led to the existence of light waves.

On the other hand, applying the quantum rules to these oscillators leads to the emergence of infinitely many discrete energy levels for each frequency, and, befitting the allowed energies of harmonic oscillators, these levels are equally spaced. Since it takes exactly as much energy to ascend from the fifth to the sixth level as from the tenth to the eleventh, the solutions of these quantized equations are readily interpreted as describing the existence of discrete energy packets, all of the same size—that is, photons: when the oscillator of frequency f is in its ground state, there are no photons. When it is on the second level, there is one photon of energy hf. When on the sixth level, there are five photons of that energy, and so on. What is more, these photons turn out to satisfy exactly the appropriate relation between their energy and momentum that the theory of relativity dictates for particles of zero mass, moving always at the speed c of light: their energy equals the magnitude of their momentum, multiplied by c. Thus, the classical treatment of the harmonic oscillators described by Maxwell's equations leads to the wave nature of light, while the quantized version of the same oscillators implies that light consists of Einstein's photons.

Electromagnetism was but the first interaction to be treated as a field, initially classically and, some seventy years later, quantum mechanically. Dirac's greatest contribution was to take an analogous step for electrons. Each of these particles was known to carry both an electric charge of exactly the same amount, which had been measured in 1913 by the American physicist Robert Andrew Millikan (1868–1953), and an intrinsic angular momentum, called *spin,* as though it constantly rotated on its own axis, making it at the same time a tiny magnet. The latter property had been discovered in 1925, but it could only be added to their quantum description in an ad hoc fashion. What was missing was a quantum-mechanical equation to describe the behavior of electrons that would take account of the fact that inside atoms as well as in many experimental contexts electrons move at speeds approaching the speed of light. This equation would therefore have to satisfy the requirements of the special theory of relativity, which those introduced earlier by Schrödinger and Heisenberg did not. Schrödinger had tried his hand at this task, but the relativistic equation he had proposed did not correctly account for experimental observations of atomic spectra.

Like a work of art, the equation proposed in 1928 by

Dirac was created on purely aesthetic grounds—constrained, of course, by the known laws of physics and the rules of mathematics. Dirac made no secret of being motivated largely by aesthetic criteria, and he succeeded marvelously: his equation is universally regarded by physicists as remarkably beautiful. It was not only an object of beauty, to be written down and gazed at with admiration, however. It also *worked.* Not only did it predict the fine details of the spectrum of the hydrogen atom, but it automatically incorporated the correct spin of the electron. In fact, when viewed from the perspective of harmonic oscillators, it accomplished even more.

Although Dirac had not looked at it in this light at first, his equation eventually had to be considered as governing a quantum field, rather than a wave function, as Schrödinger's equation had done. Just as with Maxwell's equations, Dirac's equation was most conveniently subjected to Fourier's analysis and thus transformed into a set of equations for harmonic oscillators. When "quantized," these consequently led to results analogous to those for light: the equal spacing of the energy levels of the oscillators immediately suggested an interpretation in terms of the existence of particles: this time these particles were electrons, with the correct relation between energy and momentum for particles of a given mass m as required by the theory of relativity, as well as with the right spin and the given electric charge $-e$. (That electrons were conventionally taken to carry a negative charge was a historical accident.)

Mysteriously, however, the Dirac equation not only described particles of mass m and charge $-e$ but also described particles of the same mass m and charge $+e$—yet no such objects were known to exist. It did not take long for just such particles to be discovered. The American physicist Carl David Anderson (1905–1991), in 1932, experimentally confirmed the existence of these *positrons*. Dirac's beautiful equation had hit the trifecta: it satisfied the special theory of relativity, it incorporated the spin of the electron in a natural way (it even included their magnetism with only a tiny error), and it predicted the existence of antiparticles for electrons.

The combination of the Maxwell equations and the Dirac equation, all interpreted as equations for quantum fields—the Dirac equation for the electric charges that give rise to the electromagnetic field and the Maxwell equations for the electromagnetic field responsible for the forces acting between electric charges—makes up the phenomenally successful theory of *quantum electrodynamics,* or QED, which took some twenty more years to come to full maturity. Only after crucial new contributions shortly after World War II by the American physicists Julian Schwinger (1918–1994) and Richard Feynman (1918–1988) and the Japanese physicist Sin-itiro Tomonaga (1906–1979) did QED ripen to the point of yielding numbers that could be compared with experimental results, which it did with astounding accuracy.

In contrast to equations such as Schrödinger's for a particle, these field equations describe an unlimited number of quanta that can be created and annihilated at will, subject only to the availability of sufficient energy—or even in the absence of sufficient energy, their "virtual" creation for very short periods of time—and to the law of conservation of electric charge. Indeed, after the construction of powerful accelerators, which produce colliding beams of particles of high energy, the phenomenon of *pair creation,* that is, of production of electron-positron pairs, was experimentally observed on many occasions in the laboratory. QED also predicts minute effects slightly modifying atomic spectra, as well as tiny changes in some of the properties of electrons, such as their magnetism (thus accounting for the tiny discrepancy between the magnetism of the electron as predicted by the Dirac equation and its experimentally measured value). All of these could be calculated to high precision by means of the theory, and they have been subject to experimental verification with an accuracy sometimes as high as one part in a hundred billion.

During the last half of the twentieth century, many other quantum field theories have been constructed on the model of QED, though as yet without quite its success in permitting comparisons between high-precision measurements and reliable calculations. As a result, all the known constituents of matter in the universe are now regarded as quanta of a variety of fields, and all the forces between these quanta—with the sole exception of gravity, whose quantum nature is not yet understood—are seen as manifestations of fields, which also produce quanta of their own. And all these quanta arise essentially from the same kind of mathematical mechanism, namely, the quantization of harmonic oscillators, whose energy levels are equally spaced and thus call for a particle interpretation.

The same holds for the vibrations heard as sound: they too are subject to quantum rules and thus are found to consist of particles, called *phonons.* (Since the frequencies of sound are much lower than those of light, the Planck rela-

tion $E = hf$ tells us that the energies of individual phonons tend to be much smaller than those of photons.) What
is more, even the currently fashionable theories and speculations concerning a long-awaited reconciliation between
Einstein's general theory of relativity and the quantum,
known in their various guises as *superstring theory*, employ
as their basic element the properties of Galileo's simple pendulum—like the vibrations of a harp string in higher dimensions.

The swinging pendulum has led us down a long and sin-
uous route, which began with the regulating of time by
means of stable and reliable clocks, bringing order to the
rhythms of everyday life. We moved on to the vibrations of
strings and membranes, flutes and organ pipes, which produce musical harmonies, then to light waves with interference fringes. Finally, via Einstein's photons and Dirac's electrons, we arrived at the cause of all the constituents of the
universe. Complex and forbidding though nature appears
to be, the guiding belief of most scientists is that she will
ultimately reveal herself to be both coherent and simple;
otherwise we would have scant chance of unraveling her se-
crets. Little could Galileo have realized that the harmonic
oscillator, whose isochronism he discovered in his youth,
would turn out to be the most basic, all-pervading physical
system in the world and a crucial building block for our
understanding of nature.

Notes

1. Biological Timekeeping: The Body's Rhythms

1. For a description of the mathematics involved in such synchronization phenomena, see S. H. Strogatz and I. Stewart, "Coupled oscillators and biological synchronization," *Scientific American,* December 1993, pp. 102–109. See also Steven H. Strogatz, *Sync: The Emerging Science of Spontaneous Order* (New York: Theia, 2003).

2. For a discussion of the mathematical ramifications of the cyclic nature of biological clocks, see A. T. Winfree, *The Geometry of Biological Time,* 2nd ed. (New York: Springer-Verlag, 2001).

3. I. Provencio et al., *Nature* 415 (31 January 2003): 493; S. Hattar et al., *Science* 295 (8 February 2002): 1065; D. M. Berson et al., ibid., pp. 1070, 955; see also M. Menaker, *Science* 299 (10 January 2003): 213.

4. For a review of much recent work, see S. M. Reppert and D. R. Weaver, "Coordination of circadian timing in mammals," *Nature* 418 (29 August 2002): 935–941.

5. For recent research on the circadian clock of Neurospora see C. Kramer et al., *Nature* 421 (27 February 2003): 948–952.

6. For instructions of how to perform chemical experiments with similar results at home, see the "Amateur Scientist" column by Jearl Walker, *Scientific American,* July 1978, pp. 152–158.

2. The Calendar: Different Drummers

1. If the recent speculations that Stonehenge was constructed to resemble female genitalia are correct, the purpose of that structure, of course, may have been quite different.

2. Since, according to the Bible, the resurrection of Jesus took place three days after his crucifixion, which occurred right after the Passover holiday, the date of the Easter celebration is tied to the Jewish calendar, according to which Passover is celebrated on the 14th day of the lunar month in which that day falls on or after the vernal equinox.

3. Early Clocks: Home-Made Beats

1. Quoted in R. Burlingame, *Dictator Clock: 5000 Years of Telling Time* (New York: Macmillan, 1966), p. 63.

4. The Pendulum Clock: The Beat of Nature

1. The usefulness Galileo's telescope for naval purposes was not lost on the rulers of a maritime power such as Venice, who offered him a lifetime appointment at their university; however, he declined in favor of Tuscany.

5. Successors: Ubiquitous Timekeeping

1. Since the earth's rotation is gradually being slowed down by the drag of the moon's gravitational field, the time defined by an atomic clock keeps on running slightly ahead of the "coordinated universal time," which follows a smoothed out version of the sun and thus of the earth's rotation, and which is the standard time reference for clocks around the world. To "entrain" the two, the International Telecommunication Union has, since 1972, inserted occasional "leap seconds"—by now a total of 32—in the latter.

This practice, however, has recently become controversial; see "Astronomers Leap to the Defence of Extra Seconds in Time Debate," *Nature* 423 (12 June 2003): 671.

6. Isaac Newton: The Physics of the Pendulum

1. The quotation, from a woman who used to be a childhood friend of Newton's, comes from Richard S. Westfall, *The Life of Isaac Newton* (Cambridge: Cambridge University Press, 1993), p. 13.

References

Audoin, Claude, and Bernard Guinot. *The Measurement of Time: Time, Frequency, and the Atomic Clock.* Cambridge: Cambridge University Press, 2001.

Barinaga, M. "New timepiece has a familiar ring." *Science* 281 (4 September 1998): 1429–1430.

Bedini, Silvio A. *The Pulse of Time.* Florence, Italy: Leo S. Olschki, 1991.

Berson, D.M., et al., "Phototransduction by retinal ganglion cells that set the circadian clock," *Science* 295 (8 February 2002): 1070–1073.

Brady, John, ed. *Biological Timekeeping.* Cambridge: Cambridge University Press, 1982.

Brearly, Harry C. *Time Telling through the Ages.* New York: Doubleday, Page & Co, 1919.

Breasted, James Henry. "The beginnings of time-measurement and the origins of our calendar." In *Time and Its Mysteries,* series I. New York: New York University Press, 1936, pp. 59–94.

Buijs, R. M., et al., eds. *Hypothalamic Integration of Circadian Rhythms.* Progress in Brain Research, vol. III. Amsterdam: Elsevier, 1996.

Bünning, Erwin. *The Physiological Clock: Circadian Rhythms and*

Biological Chronemetry, 3rd ed. New York: Springer Verlag, 1973.

Burlingame, Roger. *Dictator Clock: 5000 Years of Telling Time.* New York: Macmillan, 1966.

Cowan, Harrison J. *Time and Its Measurement: From the Stone Age to the Nuclear Age.* Cleveland, World Publishing Co., 1958.

Edmunds, Jr., Leland N. *Cellular and Molecular Bases of Biological Clocks.* New York: Springer-Verlag, 1988.

French, A. P., ed. *Einstein: A Centenary Volume.* Cambridge: Harvard University Press, 1979.

Goldbetter, A. *Biochemical Oscillations and Cellular Rhythms,* 2nd ed. Cambridge: Cambridge University Press, 1997.

——— "A model for circadian oscillations in the *Drosophila* period protein (PER)." *Proceedings of the Royal Society: Biological Sciences* 261 (1995): 319–324.

Golden, S. S., et al. "Cyanobacterial circadian rhythms." *Annual Review of Plant Physiology and Plant Molecular Biology* 48 (1997): 327–354.

Gonze, Didier, et al. "Robustness of circadian rhythms with respect to molecular noise." *Proceedings of the National Academy of Sciences USA* 99 (2002): 673–678.

Goudsmit, S. A., and R. Claiborne, eds. *Time.* New York: Time Inc., 1966.

Hattar, S., et al. "Melanopsin containing retinal ganglion cells: architecture, projections, and intrinsic photosensitivity," *Science* 295 (8 February 2002): 1065—1070.

Hendry, John. *James Clerk Maxwell and the Electromagnetic Field.* Bristol: Adam Hilger Ltd, 1986.

Hood, Peter. *How Time Is Measured.* Oxford: Oxford University Press, 1969.

Jacklet, J. W. "Circadian clock mechanisms." In John Brady, ed., *Biological Timekeeping.* Cambridge: Cambridge University Press, 1982, pp. 173–188.

Klarreich, Erica. "Huygens's clocks revisited." *American Scientist* 90 (2002): 322—323.

Klein, D. C., R. Y. Moore, and S. M. Reppert. *Suprachiasmatic Nucleus: The Mind's Clock*. New York: Oxford University Press, 1991.

Kluge, Manfred. "Biochemical rhythms in plants." In John Brady, ed., *Biological Timekeeping*. Cambridge: Cambridge University Press, 1982, pp. 159–172.

Leloup, J.-C., and A. Goldbetter. "Modeling circadian oscillations of the PER and TIM proteins in *Drosophila*." In Y. Touitou, ed., *Biological Clocks: Mechanisms and Applications*. Amsterdam: Elsevier, 1998, pp. 81–88.

Marcus, G. J. *A Naval History of England: I. The Formative Centuries*. Boston: Little Brown and Company, 1961.

Menaker, M. "Circadian photoreception." *Science* 299 (10 January, 2003): 213–214.

Moore, W. *Schrödinger: Life and Thought*. Cambridge: Cambridge University Press, 1989.

Moore-Ede, M. C., F. M. Sulzman, and C. A. Fuller. *The Clocks That Time Us*. Cambridge: Harvard University Press, 1982.

Needham, Joseph. *Science and Civilization in China*. Cambridge: Cambridge University Press, 1954.

Needham, J., W. Ling, and D. J. de Solla Price. *Heavenly Clockwork*. Cambridge: Cambridge University Press, 1960.

Nilsson, Martin P. *Primitive Time-Reckoning*. Lund: C. W. K. Gleerup, 1920.

Pais, A. *Niels Bohr's Times*. Oxford: Clarendon Press, 1991.

Palmer, John D. *The Living Clock: The Orchestrator of Biological Rhythms*. New York: Oxford University Press, 2002.

Pikovsky, A., M. Rosenblum, and J. Kurths. *Synchronization: A Universal Concept in Nonlinear Science*. New York: Cambridge University Press, 2002.

Reppert, S. M., and D. R. Weaver. "Coordination of circadian timing in mammals." *Nature* 418 (29 August 2002): 935–941.

Siffre, Michel. *Beyond Time*. New York: McGraw-Hill, 1964.

Shotwell, James T. "Time and historical perspective" (1946). In *Time and Its Mysteries,* Series III. New York: New York University Press, 1949, pp. 63–91.

Sobel, Dava. *Longitude*. New York: Walker and Co., 1995.

———— *Galileo's Daughter*. New York: Walker & Company, 1999.

Sobel, Dava, and William J. H. Andrewes. *The Illustrated Longitude*. London: Fourth Estate Ltd, 1998.

Strogatz, Steven, and Ian Stewart. "Coupled oscillators and biological synchronization." *Scientific American,* December 1993, pp. 102–109.

Takahashi, J. S. "The biological clock: it's all in the genes." In R. M. Buijs et al., eds., *Hypothalamic Integration of Circadian Rhythms*. Progress in Brain Research, vol. III. Amsterdam: Elsevier, 1996, pp. 5–9.

Thomas, John M. *Michael Faraday and the Royal Institution*. Bristol: Adam Hilger Ltd, 1991.

Touitou, Y., ed. *Biological Clocks: Mechanisms and Applications*. Amsterdam: Elsevier, 1998.

Watson, F. R. *Sound*. New York: John Wiley & Sons, 1935.

Westfall, Richard S. *Never at Rest*. Cambridge: Cambridge University Press, 1980.

———— *The Life of Isaac Newton*. Cambridge: Cambridge University Press, 1993.

Wever, Rütger A. *The Circadian System of Man: Results of Experiments under Temporal Isolation*. New York: Springer Verlag, 1979.

Winfree, Arthur T. *The Geometry of Biological Time,* 2nd ed. New York: Springer-Verlag, 2001.

———— *The Timing of Biological Clocks*. New York: Scientific American Library, 1987.

Illustration Credits

1. *Lindauer Bilderbogen* no.5, ed. Friedrich Boer (Jan Thorbecke Verlag). Reproduced by permission.

2. Arthur T. Winfree, *The Timing of Biological Clocks* (Scientific American Library, 1987). Reproduced by permission of Henry Holt and Company, LLC.

3. A. Goldbeter, *Biochemical Oscillations and Cellular Rhythms* (Cambridge University Press, 1997). Reproduced by permission.

5. H. J. Cowan, *Time and Its Measurement: From the Stone Age to the Nuclear Age* (The World Publishing Co., 1958).

6. Lancelot Hogben, *Science for the Citizen,* illustrated by J. F. Horrabin (W. W. Norton and Co.). Reproduced by permission.

7. J. Needham et al., *Heavenly Clockwork* (Cambridge University Press, 1960). Reproduced by permission.

8. H. J. Cowan, *Time and Its Measurement: From the Stone Age to the Nuclear Age* (The World Publishing Co., 1958).

9. Pierre Dubois, *Historie de l'Horologerie* (Administration des Moyens Ages et la Renaissance, 1849).

10. Reproduced by permission of the Smithsonian Institution.

11. Reproduced by permission of the Bridgeman Art Library International, New York.

12. Silvio Bedini, *The Pulse of Time* (Leo S. Olschki, 1991). Reproduced by permission.

13. P. Hood, *How Time Is Measured* (Oxford University Press, 1969). Reproduced by permission.

14. P. Hood, *How Time Is Measured* (Oxford University Press, 1969). Reproduced by permission.

15. Reproduced by permission of the Metropolitan Museum of Art.

16. H. J. Cowan, *Time and Its Measurement: From the Stone Age to the Nuclear Age* (The World Publishing Co., 1958).

17. Reproduced by permission of the National Maritime Museum, London.

18. P. Hood, *How Time Is Measured* (Oxford University Press, 1969). Reproduced by permission.

19. Reproduced by permission of Bulova Watch Co.

20. Reproduced by permission of the Fellows of Trinity College, Cambridge.

26. F. R. Watson, *Sound* (Wiley & Sons, 1935).

27. Reproduced by permission of The Royal Institution, London, UK/Bridgeman Art Library.

28. Reproduced by permission of Cavendish Laboratory, University of Cambridge.

30. Reproduced by permission of the Einstein Archives, Jerusalem.

32. Reproduced by permission of Mrs. R. Braunizer.

33. Reproduced by permission of the Niels Bohr Archives, Copenhagen, Denmark.

Index

INDEX

cyanobacteria, 15, 21
cycloidal suspension, 55

Darius the Great, 30
Darwin, Francis , 14
date line, 76
Davy, Humphry, 106
daylight-saving time, 76
dimensional analysis, 87
dinoflagellate, 15
Dirac, Paul Adrien Maurice, 125,
  128, 133
Drosophila, 18, 21
drums, vibrations of, 103

Eddington, Arthur, 118
Edelgestein, Reinier, 59
Eglantine, Fabre d', 28
Egyptian calendar, 29
Egyptian hours, 34
Einstein, Albert, 115ff, 118, 122,
  124, 125
electromagnetic field, 113
electromagnetic waves, 105
electromagnetism, 110
entrain, 8
equation of time, 57, 58
escapement, 40, 72
ether, 112, 115, 120

Faraday's law, 113
Faraday, Michael, 105, 110
feedback mechanism, 19
Feynman, Richard, 135
field, 107
field lines, 108

fireflies, 8
first harmonic, 102
Fizeau, Armand Hippolyte Louis,
  112
flow of time, 4, 24
folio, 43
Forel, August, 10
Foucault, Jean Bernard Léon, 92,
  112
fountain clock, 81
Fourier coefficients, 95
Fourier components, 105
Fourier, Jean Baptiste Joseph,
  94
Franklin, Benjamin, 33, 105
frequency, 91
fundamental vibration, 102
fusee, 61

Galileo Galilei, 48ff, 67, 98
gamma-rays, 114
Gemma Frisius, 59
Geneva watch making, 62
Gerbert, 40
Germain, Sophie, 103
global positioning system, 72
glycolysis, 19
GMT, 75
gnomon, 35
Golden Number, 29
Gonyaulax polyedra, 15
GPS, 72
Graham, George, 68
Greenwich Mean Time, 75
Greenwich observatory, 75
Gregorian calendar, 32

INDEX